Principles of
Applied Geophysics

Principles of
Applied Geophysics

D. S. PARASNIS

Professor of Applied Geophysics,
University of Luleå,
Luleå, Sweden

LONDON
CHAPMAN AND HALL

First published 1962
by Methuen & Co. Ltd.
Reprint 1967
Reprint 1971 published by
Chapman and Hall Ltd.
11 New Fetter Lane, London EC4P 4EE
Second edition 1972
Reprinted 1975 and 1977
Set by Santype Ltd (Coldtype Division)
Printed in Great Britain at the
University Press, Cambridge

ISBN 0 412 10980 8

Distributed in the U.S.A.
by Halsted Press, a Division
of John Wiley & Sons, Inc., New York

Preface to the Second Edition

When yet another reprint of this book was called for recently, it was decided to revise the book thoroughly because some rapid developments had taken place in the subject since the appearance of the first edition. This new edition is not only revised but enlarged; however, the object of the monograph continues to be to give a brief but fairly comprehensive survey of the principles of applied geophysics.

The first edition was characterized by a number of topics which were not accessible in any book on applied geophysics then in existence. The reader will notice that in a like manner this second edition, while preserving the frame and contents of the first edition, contains compact discussions of several topics of practical interest not to be found in any existing book devoted to the subject as a whole. These are, to mention a few of them, the thick-sheet magnetic anomaly in an arbitrary direction, the kernel function of resistivity, the auxiliary point and the Pekeris-Koefoed methods of electric interpretation, induced polarization, transient EM methods, H-mode and E-mode VLF methods, INPUT, deconvolution of seismic records and VIBROSEIS. There are now also a separate section on analytic continuation and an appendix on Fourier transforms and convolution.

The Systéme Internationale (SI) units are used throughout in this edition except in a very few instances, where data have been reproduced from original papers, probably the first geophysical book in English to adopt the SI.

In the preparation of this second edition I have benefited by many a "Nota del traductor" appended by Dr. Ernesto Orellana in his Spanish translation of the first edition.

My thanks are due to Amoco/Gas Council North Sea Group of companies and Dr. A. A. Fitch of Seismograph Service Ltd., Great Britain for providing an example of a seismic reflection cross-section, before and after deconvolution, and its geologic interpretation.

Boliden, Sweden   D. S. PARASNIS

Contents

1 · Introduction

Geophysics is the application of the principles of physics to the study of the earth. The subject includes, strictly speaking, meteorology, atmospheric electricity, or ionosphere physics but it is in the more restricted sense, namely the physics of the body of the earth, that the word geophysics will be used in this monograph. The aim of pure geophysics is to deduce the physical properties of the earth and its internal constitution from the physical phenomena associated with it, for instance the geomagnetic field, the heat flow, the propagation of seismic waves, the force of gravity, etc. On the other hand, the object of applied geophysics with which this monograph is concerned is to investigate specific, relatively small scale and shallow features which are presumed to exist within the earth's crust. Among such features may be mentioned synclines and anticlines, geological faults, salt domes, undulations of the crystalline bedrock under a cover of moraine, ore bodies, clay deposits and so on. It is now common knowledge that the investigation of such features very often has a bearing on practical problems of oil prospecting, the location of water-bearing strata, mineral exploration, highways construction and civil engineering. Often, the application of physics, in combination with geological information, is the only satisfactory way towards a solution of these problems.

The geophysical methods used in investigating the shallow features of the earth's crust vary in accordance with the physical properties of the rocks – the last word is used in the widest sense – of which these features are composed, but broadly speaking they fall into four classes. On the one hand are the *static methods* in which the distortions of a static physical field are detected and measured accurately in order to delineate the features producing them. The static field may be a natural field like the geomagnetic, the gravitational or the thermal gradient field, or it may be an artificially applied field like an electric potential gradient. On the other hand, we have the *dynamic methods* in which signals are sent in the ground, the returning signals are detected and their strengths and times of arrival are measured at

1

suitable points. In the dynamic methods the dimension of time always appears, in the appropriate field equations, directly as the time of wave arrival as in the seismic method, or indirectly as the frequency or phase difference as in the electromagnetic method. There is a further, now considerably important, class of methods which lie in between the two just mentioned. These will be called *relaxation methods*. Their feature is that the dimension of time appears in them as the time needed for a disturbed medium to return to its normal state. This class includes the overvoltage or induced polarization methods. Finally there are what we may call *integrated effect methods*, in which the detected signals are statistical averages over a given area or within a given volume. The methods using radioactivity fall in this class.

The classification of geophysical methods into ground, airborne or borehole methods refers only to the operational procedure. It has no physical significance. Many ground methods can be used in the air, under water or in boreholes as well.

In a sense, applied geophysics, excepting the seismic methods, is predominantly a science suited to flat or gently undulating terrain where the overburden is relatively thin. The reason is that whenever the relief is violent, the data of geophysical methods need corrections which are frequently such as to render their interpretation uncertain. On the other hand, when the overburden is too thick the effects produced by the features concealed under it generally lie within the errors of measurement and are difficult to ascertain. There is, however, no general rule as to the suitability of any terrain to geophysical methods and every case must be considered carefully on its own merits.

The various methods of applied geophysics will be dealt with in turn in the following chapters.

2 · Magnetic Methods

2.1 Short history

It was in the year 1600 when William Gilbert, physician to Queen Elizabeth I, published his book *De Magnete*, that the concept arose of a general geomagnetic field with a definite orientation at each point on the surface of the earth. In its wake, observations of the local anomalies in the orientation of the geomagnetic field were used in Sweden for iron-ore prospecting, for the first time probably as early as 1640 and regularly by the end of that century. They constitute the first systematic utilization of a physical property for locating specific, small-scale features within the earth's crust. Two centuries later, in 1870, Thalén and Tiberg constructed their magnetometer for comparatively rapid and accurate relative determinations of the horizontal force, the vertical force and the declination, by the familiar sine and tangent methods used in elementary physics courses. It was in widespread use, especially in Sweden, as a tool for prospecting surveys for more than the following half a century. The large-scale use of magnetic measurements for investigations of geological structures, other than those associated with magnetite ore, did not however begin seriously until 1915, when Adolf Schmidt constructed his precision vertical field balance using a magnetic needle swinging on an agate knife edge. Since then magnetic observations have been successfully employed not only in the search for magnetite ore but also in locating buried hills, geological faults, intrusions of igneous rocks, salt domes associated with oil fields, concealed meteorites and buried magnetic objects such as pipe-lines.

2.2 Magnetic properties of rocks

The magnetic method of applied geophysics depends upon measuring accurately the anomalies of the local geomagnetic field produced by the variations in the intensity of magnetization in rock formations. The magnetization of rocks is due partly to induction in the earth's field and partly to their permanent (remanent) magnetization. The induced intensity

3

depends primarily upon the magnetic susceptibility and the magnetizing field, and the permanent intensity upon the geological history of the rock.

Susceptibility of rocks.

In accordance with the general classification used in physics rocks fall into three categories, namely diamagnetic, paramagnetic and ferromagnetic. In view of the present state of our knowledge the last named category must be subdivided into the truly ferromagnetic, the antiferromagnetic and the ferrimagnetic substances. These terms are briefly explained below.

The intensity of magnetization, M_i, induced in isotropic substances due to a magnetizing force H can be written as:

$$M_i = \kappa H \quad (\kappa = \text{volume susceptibility}) \tag{2.1}$$

Both M_i and H are vectors. In a diamagnetic body κ is negative so that the induced intensity is in a direction opposite to the magnetizing field. The origin of diamagnetism lies in the motion of an electron round a nucleus. This motion constitutes a miniature current circuit whose magnetic moment vector begins to precess round an applied external field in accordance with Larmor's well-known theorem. The additional periodic motion of the electron produces a magnetic moment opposite in direction to the applied field. It will be realized that there is a diamagnetic effect in all substances including the 'typical' ferromagnetics like iron, cobalt and nickel. But net diamagnetism only appears if the magnetic moments of atoms are zero in the absence of an external field, as is the case for atoms or ions having closed electronic shells. There are many rocks and minerals which show net diamagnetism. Chief among them are quartz, marble, graphite, rock salt, and anhydride (gypsum).

The susceptibility of paramagnetic substances is positive and decreases inversely as the absolute temperature (Curie-Weiss law). Paramagnetism makes its appearance when the atoms or molecules of a substance have a magnetic moment in the absence of a field and the magnetic interaction between the atoms is weak. Normally the moments are distributed randomly, but on the application of the field they tend to align themselves in the direction of the field, the tendency being resisted by thermal agitation. The paramagnetism of elements is mainly due to the unbalanced spin magnetic moments of the electrons in unfilled shells, like the 3d-shells of the elements from Sc to Mn. Many rocks are reported to be paramagnetic, for instance gneisses, pegmatites, dolomites, syenites, etc. However, it seems certain that their paramagnetism is not intrinsic but is a manifestation of a weak ferrimagnetism due to varying amounts of magnetite or ilmenite, or an antiferromagnetism due to minerals like haematite, manganese dioxide, etc.

In ferromagnetic materials the atoms have a magnetic moment and the interaction between neighbouring atoms is so strong that the moments of all atoms within a region, called domain, align themselves in the same direction even in the absence of an external field. In Fe, Co and Ni this interaction takes place between the uncompensated spins in the unfilled 3d-shells of the atoms. A state of spontaneous magnetization can therefore exist consisting of an orderly arrangement of the magnetic moments of all atoms. Typical of the ferromagnetics are their hysteresis loops and their large susceptibilities which depend upon the magnetizing force. Ferromagnetism disappears above a temperature known as the Curie temperature. There are no truly ferromagnetic rocks or rock minerals.

There exist substances in which the susceptibility has an order of magnitude characteristic of a paramagnetic (10^{-5} SI) but is not inversely proportional to the temperature. Instead it first increases with temperature, reaches a maximum at a certain temperature, also called the Curie point or the λ-point and decreases thereafter according to the Curie-Weiss law. In these substances the low magnetic susceptibility below the λ-point can be explained by assuming an *ordered* state of atoms such that the magnetic moments of neighbouring atoms are equal but directed antiparallel to each other. Thus the two ordered sub-lattices, each reminiscent of the state in a ferromagnetic, cancel each other and their net magnetic moment is zero. This state is called antiferromagnetism and can be confirmed by neutron diffraction studies. Of the rock-forming minerals, haematite (Fe_2O_3) is the most important antiferromagnetic (λ-point 675°C).

Among the antiferromagnetic substances there is a class in which, to put it simply, two sub-lattices with metallic ions having magnetic moments are ordered antiparallel as above, but in which the moments of the lattices are unequal, giving rise to a net magnetic moment in the absence of a field. Such substances are called ferrimagnetic. Practically all the constituents giving a high magnetization to rocks are ferrimagnetic, chief among them being magnetite (Fe_3O_4), titanomagnetite ($FeO(Fe, Ti)_2O_3$) and ilmenite

(FeTiO$_3$). Spontaneous magnetization and a relatively high susceptibility can also exist in an antiferromagnetic if statistically systematic defects are present, as is believed to be the case for pyrrhotite (FeS). The temperature dependence of ferrimagnetics is complex, there being theoretically several possibilities.

TABLE 1

Volume susceptibilities
(in rationalized units of 10^{-6})

Graphite	−100	Gabbro	3,800−90,000
Quartz	−15.1	Dolomite	
Anhydrite	−14.1	(impure)	20,000
Rock salt	−10.3	Pyrite	
Marble	−9.4	(pure)	35−60
Dolomite		Pyrite	
(pure)	−12.5 − +44	(ore)	100−5,000
Granite		Pyrrhotite	10^3-10^5
(without magnetite)	10−65	Haematite (ore)	420−10,000
Granite		Ilmenite (ore)	$3 \times 10^5 - 4 \times 10^6$
(with magnetite)	25−50,000	Magnetite (ore)	$7 \times 10^4 - 14 \times 10^6$
Basalt	1,500−25,000	Magnetite (pure)	1.5×10^7
Pegmatite	3,000−75,000		

To convert the above values to unrationalized em cgs units divide by 4π.

The susceptibility of rocks is almost entirely controlled by the amount of ferrimagnetic minerals in them, their grain size, mode of distribution, etc. and is extremely variable. The values listed in Table 1 should nevertheless serve to give a rough idea. Various attempts have been made to represent the dependence of susceptibility on the content of ferrimagnetics, but no simple universally valid relation exists. For particular groups of rocks or for particular ranges of susceptibility a statistically significant correlation can generally be found between the amount of Fe$_3$O$_4$ and the susceptibility (Fig. 1). However, the scatter is usually such that a prediction based on such correlations must be used with caution.

It will be seen that generally speaking only small susceptibility differences ($\Delta\kappa$) will be encountered between rock formations. The maximum $\Delta\kappa$ (when a deposit of high grade magnetite ore is present) is of the order of ten

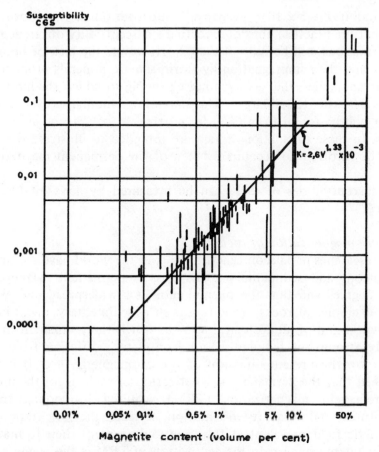

Fig. 1. Variation of susceptibility with magnetite content. After
Balsley and Buddington (Economic Geology, *Nov. 1958*)

(SI). If two very extensive and thick homogeneous formations are
separated along a plane vertical contact the total change in the vertical
magnetic field in traversing the contact is given by

$$\Delta Z = \frac{1}{2} \Delta \kappa Z \qquad (2.2)$$

if Z is the vertical magnetic field of the earth. The practical limit of
determining ΔZ in field surveys is about 1 gamma ($=10^{-9}$ weber/m^2 $= 10^{-5}$

gauss) and if $Z = 5 \times 10^{-5}$ weber/m^2, substitution in the above equation shows that the practical limit of detecting a susceptibility difference between rock formations will be $\Delta\kappa \approx 4 \times 10^{-5}$. However, the lack of homogeneity in rocks and their impregnation by ferrimagnetic minerals produce random magnetic anomalies, the 'geologic noise', owing to which the limit in reality is some 10–100 times larger. The maximum anomaly (over a magnetic deposit) will be found from (2.2) by putting $\Delta\kappa = 12$ to be about 300,000 gammas. In these calculations which are intended to illustrate the orders of magnitude involved no account is taken of the permanent magnetization of rocks.

The susceptibilities of rocks can be measured by any suitable standard method (*1*)*.

Permanent magnetization of rocks.
Recent researches in palaeomagnetism all over the world have confirmed that both igneous and sedimentary rocks possess permanent magnetization in varying degrees and that the phenomenon is a widespread one. Well-documented examples of rocks, igneous as well as sedimentary, occur in all parts of the world and of all geological ages, in which the permanent intensity is not only strong but has a direction completely different from, at times opposite to, the present direction of the geomagnetic field. It is now well established that the permanent magnetization of rocks is for the most part a thermoremanent magnetization (TRM) acquired in cooling from high temperatures and that its orientation reflects the orientation of the geomagnetic field prevalent at the time and place of their formation. The predominant mechanism in the acquisition of TRM is the alignment of the domains in the ferrimagnetic constituents of the rocks. It is in this respect significant that the TRM of rocks disappears when they are heated above 600°C which is approximately the Curie point of magnetite (*3*).

In Table 2 are given some examples of the ratio $Q_n = M_n/\kappa F$ in rocks, where M_n is the permanent intensity, and F the total magnetizing force of the earth's field. In most igneous rocks like lavas and dikes the permanent intensity completely dominates the intensity induced by the earth's field. Hence magnetic interpretation, especially in areas where igneous rocks occur,

* Italicized numbers in brackets refer to literature
references in the list at the end of the book.

TABLE 2

Q_n for some rock specimens

Specimen	Locality	Q_n
Basalt	Mihare volcano, Japan	99—118
Gabbro	Cuillin Hills, Scotland	29
Gabbro	Småland, Sweden	9.5
Andesite	Taga, Japan	4.9
Granite	Madagascar	0.3—10
Quartz dolerite	Whin sill, England	2—2.9
Diabase	Astano Ticino, Switzerland	1.5
Tholeiite dikes	England	0.6—1.6
Dolerite	Sutherland, Scotland	0.48—0.51
Magnetite ore	Sweden	1—10
Manganese ore	India	1—5

Sediments generally low.

must take into account the permanent intensity if a satisfactory geological picture is to be obtained. The same applies when the object of investigation is magnetite ore.

The resultant magnetization **M** of a rock can be described by the vector equation

$$\mathbf{M} = \mathbf{M}_n + \kappa \mathbf{F} \qquad (2.3)$$

In the two simplest cases where $\mathbf{M}_n$ and $\mathbf{F}$ are parallel and antiparallel, an apparent susceptibility

$$\kappa' = \kappa \pm M_n/F \qquad (2.4)$$

can be defined and used during interpretation in the equations specifying the anomaly of an anomalous feature. For the modification necessary in the general case reference may be made to a paper by Green (2).

2.3 The geomagnetic field

In order to identify the anomalies in the earth's field it is clearly essential to know its undisturbed character. To a very close approximation the regular geomagnetic field can be represented formally as the field of a dipole situated at the centre of the earth with its magnetic moment pointing towards the earth's geographical south. Physically, the origin of the field seems to be a system of electric currents within the earth. The total magnetic field of the earth has a flux density (on the surface) of about 0.6×10^{-4} wb/m^2. This flux density is the result of a magnetizing force existing at the point of observation of magnitude $0.6 \times 10^{-4}/(4\pi \times 10^{-7}) = 47.8$ amp/m. The vertical component of the geomagnetic flux density at, for example, Boliden, Sweden is 0.494×10^{-4} wb/m^2 ($= 0.494$ gauss $= 49,400$ gamma) and the associated magnetizing force is 39.4 amp/m. The factor $4\pi \times 10^{-7}$ represents the magnetic permeability of vacuum. At any point on the earth's **surface the magnetic flux density vector is completely specified by its** horizontal (H) and vertical (Z) components and the declination (D), west or east from the true north, of H. Z is reckoned positive if it points downwards as in the northern hemisphere generally and negative if it points upwards as in the southern hemisphere. The inclination (I), which is of importance in the interpretation of magnetic anomalies, is given by $\tan^{-1} Z/H$. The points on the earth at which $I = \pm 90°$ are called the magnetic north and south dip poles respectively. There may be any number of such points due to local disturbances which it is the purpose of applied geophysics to locate, but apart from them there are two main north and south dip poles situated approximately at 72° N, 102° W and 68° S, 146° E. On account of the irregular part of the earth's field they do not correspond to the intersections of the axis of the imaginary dipole at the earth's centre with the surface which are at 79° N, 70° W and 79° S, 110° E. The latter are called the geomagnetic poles or axis poles. It will be seen that the geomagnetic north pole is in reality situated at the geographical south and the geomagnetic south pole at the geographical north. But, for convenience, the two magnetic poles are often called by the name of the nearest geographical pole. It is important to keep this practical usage in mind. The imaginary line on the earth's surface passing through the points at which $I = 0$ is called the magnetic equator. North of it Z is positive, south of it, it is negative.

The earth's field is not constant at any point on its surface but undergoes variations of different periods. From the standpoint of applied geophysics the most important are the diurnal variations and magnetic storms. Their disturbing effect must be suitably eliminated from magnetic survey observations.

Further details about the geomagnetic field may be found in another monograph, *The Earth's Magnetism* (Methuen and Co.) by S. Chapman.

TABLE 3

Values of *H, Z, D* (1965)

Place	Lat.	Long. East	H Gauss	Z Gauss	D East
Chelyuskin	+77.7°	104.3°	0.0337	0.5890	+20.7°
Tromsö	+69.7	18.9	0.1125	0.5109	0.0
Hartland	+51.0	355.5	0.1887	0.4354	-9.5
San Miguel	+37.8	334.4	0.2498	0.3849	-15.0
Hyderabad	+17.4	78.6	0.4014	0.1500	-1.7
Tangerang	-6.2	106.6	0.3721	-0.2388	+1.1
Mauritius	-20.1	57.6	0.2222	-0.3049	-17.1
Gnangare	-31.8	116.0	0.2391	-0.5350	-2.9
Macquarie Is.	-54.5	159.0	0.1315	-0.6421	+26.5

(1 gauss = 10^5 gamma = 10^{-4} weber/m^2)

Values of *H, Z, D* for the epoch 1965 at some selected places are given in Table 3.

2.4 Instruments of magnetic surveying

Magnetic measurements in applied geophysics are most conveniently carried out as relative determinations in which the values of one or more elements of the magnetic field at any point are expressed as differences from the values at a suitably chosen base point. Usually the area of investigation is relatively small, say a few square kilometres, so that the normal geomagnetic field within it may be considered to be substantially constant and equal to that at the base point. In very large areas, say more than a hundred square kilometres, the variation of the normal field may be significant, especially in the north–south direction, and should be corrected for.

There exists a variety of instruments suitable for relative magnetic determinations. The principal types are described below.

2.4.1 *Pivoted needle instruments.*

The sensitive element in these is a magnetic needle arranged to swing on pivots. Probably the oldest known instrument of this type is the *Swedish mine compass* in which a compass needle can rotate in the horizontal as well as the vertical plane and take a position along the total intensity vector. In the *Hotchkiss superdip* the total intensity is determined from the deflection of a system consisting of a magnetic needle to which is fastened a counterarm carrying a small weight and making an adjustable angle with the needle. The system is suspended on a horizontal axle in the magnetic meridian. In the *Thalén-Tiberg magnetometer*, a magnetic needle with an adjustable counterweight is suspended on hardened steel bearings in a small glass-covered case which can be swung on a horizontal axle and held in the vertical or the horizontal plane. The case is first clamped in the horizontal position so that the needle swings on a vertical axis. The direction of the local H having been noted from the equilibrium position of the free needle, its magnitude is determined by means of an auxiliary magnet of known moment with the well-known Lamont sine method. The glass case is then held in a plane perpendicular to the magnetic meridian and the inclination of the needle, which is not vertical owing to the counterweight, is used in combination with the calibration constant to calculate Z.

2.4.2 *Schmidt-type variometers.*
In the *Schmidt variometer* designed to measure Z, a magnetic system is free to swing on an agate knife edge in a vertical plane like the beam of a weighing machine. Its equilibrium position at the reference station is adjusted (by altering the centre of gravity) to be horizontal and the deflections from this position at other stations are read by means of an auto-collimating telescope. On multiplying by the calibration constant they yield the relative values of Z in appropriate units. In order to eliminate the effect of H, the magnetic system is always orientated at right angles to the magnetic meridian when taking observations. The variometer designed to measure H is similar except that the magnet system is suspended initially in a vertical position and readings are made in the magnetic meridian. Small errors in orientation affect the readings of the horizontal force variometers much more than the vertical ones.

2.4.3 *Compensation variometers.*
These instruments are similar to the Schmidt variometers, but instead of measuring the tilt of the magnetic system from the horizontal they measure the force needed to restore it to that position. The magnetic needle generally

hangs on thin wires instead of being balanced on knife edges, and the restoring force is obtained by turning or moving compensating magnets, as in the various *ABEM* models, or by means of a torsion head, as in the *Askania* type. Since the deflection moment due to H is zero when the needle is horizontal the compensation-type variometers can be used in any azimuth.

2.4.4 *Flux-gate instruments*.

In these instruments use is made of the fact that magnetic fields as small as the earth's field induce in certain high permeability materials flux densities which are an appreciable fraction of their saturation densities. The a.c. wave form derived from the coils wound on cores of such materials is distorted if a steady magnetic field is superimposed on them. By means of suitable amplifying and rectifying circuits this distortion can be made to exhibit the steady field, as the deflection of a meter or a recording pen.

2.4.5 *Proton free-precession magnetometer*.

This instrument utilizes the phenomenon of nuclear magnetic resonance. A strong magnetic field is applied to a bottle of water by a coil wound round it. The protons, on account of their spin moments, align themselves parallel to this field. As soon as it is removed the spins begin to precess round the direction of the prevalent earth's field in accordance with Larmor's theorem and induce a small voltage in the coil. The frequency of this voltage which can be measured is γT, where γ is the gyromagnetic ratio of the protons and T the earth's total field (flux density). One advantage of this magnetometer is that it requires no levelling.

The accuracy of pivoted needle instruments is scarcely better than ± 100 gammas and they are now seldom used. The Schmidt and compensation type variometers are precision instruments having accuracies better than ± 5 gammas, although the accuracy of some instruments of this type, intended for rough reconnaissance surveys, is not better than $\pm 20-50$ gammas. The flux-gate and proton free-precession magnetometers have the advantage that their sensitive elements and the measuring or the recording system can be widely separated by cables. They can, therefore, be conveniently used under certain circumstances (e.g. in boreholes, under water, in airborne work, etc.) where the design of conventional instruments would be highly complicated. A precision of ± 10 gammas is possible in both these magnetometers.

The speed of modern magnetometers can be gathered from the fact that one observation with them takes only about 45 to 60 seconds.

2.5 The measurement of magnetic anomalies

Among the magnetic elements, the direction of the field is the element least sensitive to changes in the dimensions and magnetic properties of a subsurface body. It is therefore never used by itself in accurate work. Of the remaining, namely H, Z and T, any one or more can be chosen for measuring the respective anomalies ΔH, ΔZ and ΔT. Since ΔT is associated with a change also in the direction of T, the interpretation of these anomalies becomes somewhat complicated. Total field determinations are seldom resorted to in ground work. They are, however, utilized frequently in airborne and borehole work on account of instrumental advantages. Between the ΔH and ΔZ components, measurements of the latter are to be generally preferred since the subsurface picture can be visualized much more readily from ΔZ than from ΔH. Sometimes both of them are measured. In a survey over an area which may be considered to be virtually an infinite plane, such a procedure offers no additional information if observations of only the one component are made sufficiently densely, and is therefore superfluous. This rather surprising result follows from Green's theorem in potential theory and will be proved in the chapter on gravitational methods. However, in certain special cases (e.g. in underground work) where the measurements cannot be carried out over a plane (or a closed surface) or where a sufficiently dense network of points is not available, the knowledge of both ΔH and ΔZ can be of some additional help.

Another surprising result of practical importance is that the total change in ΔZ across geological features such as thin dikes or spherically shaped masses is always (that is, even in low magnetic latitudes) greater than or equal to that in ΔH (cf. Fig. 5 and 6). Thus, from this point of view too, measurements of ΔZ are to be preferred to those of ΔH *all over the world.*

The discussion in this chapter will be confined, in general, to the vertical component and its anomalies but will be extended to the other components as occasion demands.

2.6 Field procedure

When the area for magnetic investigation has been selected a base line is staked parallel to the geological strike (the trace of the rock strata on the plane represented by the surface of the earth) and measurements are made at regular intervals along lines perpendicular to the base. A reference point, far from artificial disturbances such as those due to power lines, railways, etc. is chosen and the magnetic intensity values at all other points are measured as positive or negative differences from the intensity at this point. The reference point may, strictly speaking, be anywhere within or without the area, but it is convenient to choose a point, as near as possible to the area or inside it, at which the magnetic field is known to be approximately the normal field.

Certain precautions must be taken in magnetic measurements. Magnetic materials in the wearing apparel of the observer, like keys, penknives, wrist-watches (also 'non-magnetic' ones!), etc. can totally vitiate the observations and ought to be entirely removed. Observations in the neighbourhood of iron objects such as motor cars, iron-junk, railroads, bridges should be avoided.

Several corrections must generally be applied to a set of magnetometer observations as follows: (1) To correct for the diurnal variations of the earth's field, an auxiliary instrument is kept at some convenient station within the area and read at intervals. Automatic recording may also be used. The algebraic difference between the readings of this instrument at any time, t, and the reference time, $t = 0$ (corrected, if necessary for the temperature coefficient) is subtracted from the reading at the field station, which was also measured at time t. (2) Corrections for the temperature dependence of the field instrument can be applied if its temperature coefficient is known. Most modern instruments are, however, temperature compensated and the correction is therefore negligible. (3) Terrain corrections (4): Rough terrain may give rise to anomalies. For instance, if the rocks above a station situated in a depression are magnetic, 'false' negative anomalies will be recorded. There are no general rules for applying terrain corrections. Usually, the anomalies showing a strong correlation with the terrain are regarded as less significant than others.

If great accuracy is not desired, it is sufficient to repeat at 1–2 hours' interval the reading at a previously occupied station and distribute the difference over the stations measured during this interval. The correction thus obtained includes both the diurnal variation and the temperature coefficient of the instrument.

An important point in considering the anomalies in an area is the zero level, that is the readings of the instrument at points where the field is the normal undisturbed geomagnetic field. If the readings remain constant, or vary randomly so as to suggest a geologic noise only, over a sufficient length of a measuring profile, say 100–500 m, the reading at any point on this stretch may be taken to be the zero reading and the anomalies at all other points in the area referred to it. If distinct magnetically anomalous masses are evident in the area, the zero level can be determined from the flanks of an anomaly curve, since they approach it asymptotically at great distances from the mass. Regional anomaly gradients, terrain characteristics, or contacts between rock formations of differing magnetizations, sometimes make it impossible to use the same zero level throughout the area.

2.7 The interpretation of magnetic anomalies

The magnetic anomaly field A satisfies Laplace's equation in potential theory, namely

$$\nabla^2 A = 0 \qquad (2.5)$$

This equation does not suffice to find the subsurface magnetization (see further, Chapter 3). The usual procedure in interpreting magnetic (or gravity) anomalies is to guess a body of suitable form, calculate its field on the surface and compare it with observations. It is then possible to adjust the depth and dimensional parameters of the body by trial and error until a satisfactory fit is obtained. Such a solution is only one of an infinity of possible solutions. The essential assumption in magnetic claculations is that of homogeneous magnetization which is, however, only valid for bodies bounded by second degree surface (spheres, ellipsoids, infinitely long cylinders, etc.). Geologic structures do not conform to these shapes except in very rare instances and even then only approximately. At points inside a geologic body far from its bounding surface, the above assumption may be more or less true, but at the edges and corners it definitely fails. Consequently, all calculations of the magnetic field of a geologic structure are approximations of varying degree which conform better to the true field the greater the distance from the structure.

A first step towards interpretation is the preparation of a 'magnetic map' on which the intensity values at different stations are plotted and on which the contours of equal ΔZ (isoanomalies) are drawn at suitable intervals. The interpolation necessary in contour drawing is particularly easy if the observation points form a square network. In such cases, the most accurate contours are obtained if the interpolation is carried out mainly along the approximate direction of the isoanomalies, since this is the direction in which the gradients are low. A trial isoanomalous line is first sketched to obtain the trend and is subsequently corrected by more exact interpolation.

Certain qualitative conclusions are readily drawn from a magnetic map. Anomalous conditions in the subsurface are indicated, for instance, by successive closed contours with the ΔZ-values increasing or decreasing towards a 'centre' while the direction of elongation of the closed curves may be identified with the strike of the anomalous body. Another indication is given by high horizontal anomaly gradients. They are often associated with contacts between rocks of different susceptibilities or unequal total intensities of magnetization, the contact lying shallower the steeper the gradient.

As regards strike it should be noted that in low magnetic latitudes the above rule of thumb should be used with great caution, because in these regions, bodies of finite length, striking north-south, produce anomaly patterns which indicate an apparent east-west strike. The effect is easily understood by considering the poles developed at the north and south extremities of the body and by noting the near-absence of poles on the upper surface. In fact, the anomaly pattern in this case closely resembles that in Fig. 4 discussed later. For strike directions of the body deviating from the north too, the anomaly patterns in low magnetic latitudes indicate apparent strikes significantly different from true ones (*13*). In high magnetic latitudes the effect of a strong, almost horizontal remanent magnetization may similarly cause the anomaly patterns to deviate from the true strike.

Despite the non-uniqueness of the solutions of equation (2.5) the interpretation of magnetic anomalies is not so seriously hampered in practice as might be imagined. The reason is that on the basis of geological information, which is usually available, or on grounds of plausibility, it is normally possible to reduce the number of alternatives to a moderate one from which only a few need be selected initially as working hypotheses for the trial calculations.

The anomalies of line poles.

The simplest approach to quantitative interpretation is to assume that each anomaly centre corresponds to a single magnetic pole of appropriate length and strength. The magnetization of a thin, vertical (or steeply dipping) dike, or ore sheet of great depth extent, can be approximated thus. In the two extreme cases, when the body is very short and very long compared to the depth, the anomalies reduce to those of a point pole and an infinite horizontal line pole. Along a profile over the pole, and in the second case also perpendicular to it, they are readily shown to be (in wb/m²)

$$\Delta Z = \frac{\mu_0}{4\pi} \frac{m}{a^2} \frac{1}{(1 + x_a^2)^{3/2}} \quad \text{(point pole of strength } m) \qquad (2.6)$$

$$\Delta Z = \frac{\mu_0}{4\pi} \frac{2\lambda}{a} \frac{1}{(1 + x_a^2)} \quad \text{(line pole of strength } \lambda \text{ per m)} \qquad (2.7)$$

where x = horizontal distance from an origin vertically above the pole, a is the pole depth, μ_0 is the permeability of vacuum ($4\pi \times 10^{-7}$ henry/m) and

$$x_a = x/a$$

The anomaly along an arbitrary profile across a line pole of finite length is given by $\Delta Z = \Delta Z_\infty [f(d_{1a}) + f(d_{2a})]$ where $d_{1a} = d_1/a$ and $d_{2a} = d_2/a$ are the distances of the profile from the two ends of the pole and $f(d_a) = d_a/(x_a^2 + d_a^2 + 1)^{\frac{1}{2}}$.

Now, from the definition (pole strength per unit area) of the intensity of magnetization $\kappa Z/\mu_0$, it is evident that $m = A\kappa Z/\mu_0$ where A is the horizontal cross-sectional area of the point pole and $\lambda = b\kappa Z/\mu_0$, where b is the width of the thin, infinitely long plate. Then it is easy to show the following relations with the help of (2.6) and (2.7):

Point pole

$$\left. \begin{array}{l} \Delta Z_{max} = A\kappa Z/4\pi a^2 \\ \\ a = (2^{2/3} - 1)^{-1/2} x_{1/2} = 1 \cdot 305 \, x_{1/2} \end{array} \right\} \qquad (2.8)$$

Infinite line pole

$$\left. \begin{array}{l} \Delta Z_{max} = 2b\kappa Z/4\pi a \\ \\ a = x_{1/2} \end{array} \right\} \qquad (2.9)$$

where $x_{1/2}$ is the distance at which $\Delta Z = \frac{1}{2}\Delta Z_{max}$. The depth and the parameters $A\kappa$ and $b\kappa$ may be calculated from (2.8) and (2.9).

Equations corresponding to (2.6) and (2.7) can be derived also for the general case of a pole of length L (5). If we define a distance $y_{1/2}$ (analogous to $x_{1/2}$) along a line parallel to and directly above such a pole, the ratio $x_{1/2} : y_{y2}$ depends on L/a and has the limiting values 1.000 (point pole) and 0 (infinite line pole). Similarly $a/x_{1/2}$ will be a function of L/a with the limiting values 1.305 and 1.000 respectively. In Fig. 2 are plotted $x_{1/2} : y_{1/2}$, $a/x_{1/2}$ and the quantity $b\kappa Z/4\pi x_{1/2}\Delta Z_{max}$ (equal from (2.9) to $\frac{1}{2}$ for an infinite line pole), against L/a. If $x_{1/2} : y_{1/2}$ can be determined from the observed anomaly profiles, the depth to the upper surface, length and 'magnetic width' ($b\kappa$) of a body which can be approximated by a thin,

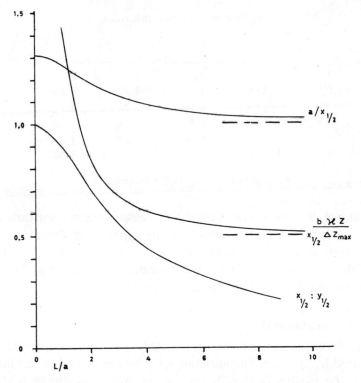

Fig. 2.   *Line pole parameters as a function of pole length*
(The parameter of the middle curve should have 4π in the denominator
if κ is expressed in SI.)

vertically magnetized, steeply dipping plate of very large depth extent, may be estimated from Fig. 2. The transverse magnetization of the plate is neglected in these calculations.

If it is desired to take into account the effect of a finite depth extent, ΔZ may be calculated from equations (2.6) and (2.7) by subtracting from their right-hand sides expressions of identical forms but with a', the depth of the lower surface, replacing a. Some examples of anomaly curves of such magnetic 'double lines' are shown in Fig. 3. The positions of the zero and the negative minimum through which ΔZ passes in these cases depends on L/a and a'/a and can be used to estimate the depth extent. The ratio $a/x_{1/2}$ depends likwise on L/a and a'/a (Table 4) so that the depth of the upper

TABLE 4

$a/x_{1/2}$ for magnetic double lines

L/a	\multicolumn{4}{c}{a'/a}
	1	2	4	∞
0	1.99†	1.54	1.37	1.305
2	1.91‡	1.45	1.26	1.18
4	1.88‡	1.43	1.21	1.08
8	2.02‡	1.48	1.20	1.03
∞	2.06‡	1.53	1.26	1.00

† Point dipole.   ‡ Linear dipoles.

surface cannot be calculated without a reasonable assumption about the depth of the lower one. The effect of an appreciable deviation of the dip from the vertical may be taken into account by shifting the lower pole sideways. In that case the negative minima in Fig. 3 will be greater in magnitude on the dip side than on the opposite one.

2.8 Spheres and ellipsoids

Bodies whose linear dimensions are roughly equal in all directions may be approximated by spheres. The anomaly of a sphere placed in the magnetizing force F of the earth's field is the same as that of a dipole with moment (amp$-$m^2)

$$m = \frac{1}{3}b^3 \frac{\kappa}{1 + \frac{1}{3}\kappa} F \tag{2.10}$$

(b = radius of the sphere) placed at the centre of the sphere. It can be shown to be

$$\Delta Z = \frac{C}{r^3}\left(\frac{3a^2}{r^2} - \frac{3 \cot I \, \cos\alpha ax}{r^2} - 1\right) \tag{2.11}$$

where $C = \frac{1}{3}b^3 \kappa Z/(1 + \frac{1}{3}\kappa)$,

x = coordinate along the earth's surface measured from an origin vertically above the sphere's centre,

a = depth of centre below the surface,

$r^2 = a^2 + x^2$ and

α = angle between the magnetic meridian and the line Ox.

x is reckoned positive in the direction of the component of the earth's horizontal field along Ox. Rules for determining the depth to the centre can be formulated in a manner similar to equations (2.8) and (2.9). As may be

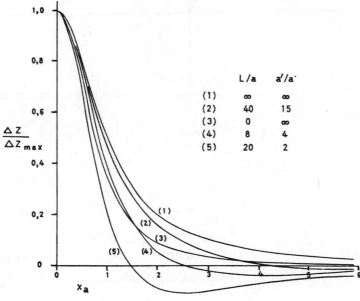

Fig. 3. ΔZ for line pole systems

expected $a/x_{1/2}$ differs but little from the value 1.99 in Table 4 for $L/a = 0$ and $a'/a = 1$ and is identical to it when $\alpha = 90°$.

The lines of equal ΔZ and, for comparison, also those of ΔH due to a sphere just touching the surface of the earth ($a = b$) at a place where $Z = 45,000$ gammas and $H = 20,000$ gammas ($I = 66°$) are shown in Fig. 4. In Fig. 5 are reproduced the profiles of ΔZ and ΔH along the lines of measurement passing through the origin and lying in the magnetic meridian ($\alpha = 0°$) over spheres with different directions of magnetization. The abscissae are in units of a. It will be seen that north of the magnetic equator ($I > 0$), the maximum positive ΔZ as well as ΔH occurs south of the point vertically above the centre of the sphere. The conditions south of the magnetic equator are reverse. In both cases, the displacement of the absolute maximum anomaly in ΔZ and ΔH from the origin is greater the smaller the inclination I.

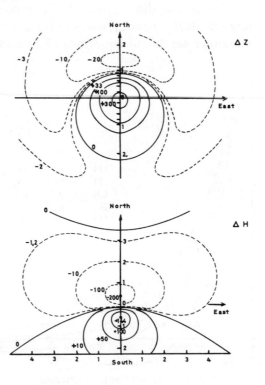

Fig. 4.   *Contours of ΔZ and ΔH above a magnetized sphere. After Haalck (Ref. 15)*

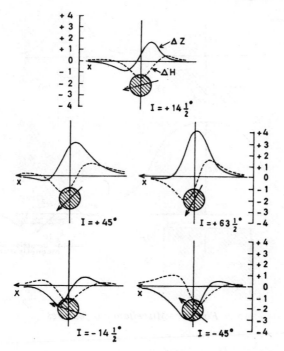

Fig. 5.   ΔZ and ΔH profiles across a sphere. After Haalck
(Ref. 15)

Exact expressions similar to (2.11) may also be obtained for infinitely long cylinders and ellipsoids (6, 7). For ellipsoids, however, they are very complicated. But in both these cases the anomalies are similar in their general features to those of a sphere. Of course, geologic bodies cannot be expected to assume these simple shapes, but curves such as those in Fig. 4 and 5 are of considerable help in qualitative interpretation.

2.9 Miscellaneous features

Some further geological features of importance are shown in Fig. 6. The step structure in (a), assumed to be very long perpendicular to the plane of the paper, is often encountered as a fault when rock beds have been thrust up or down with respect to each other along a slip-plane. It may also represent the

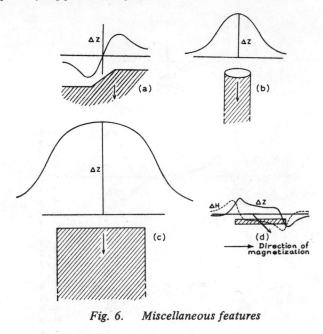

Fig. 6.   Miscellaneous features

undulations of the bedrock under an overburden, a flexure of rock beds or an erosional mountain-valley combination. The cylinder in (b) may be a salt dome or an intrusion. The thick plate in (c) may approximate to a broad zone of magnetically impregnated rock or a thick dike, while the plate in (d) may be encountered as a horizontally-lying feature. Typical magnetic anomalies over them are also shown. They are calculated on the assumption of homogeneous magnetization.

Of the various features shown, (a) and (c) merit further attention on account of their widespread occurrence geologically. We shall start with (c).

Thick sheet.

Fig. 7a shows such a sheet in plan and in a vertical section at right angles to the length. The magnetization M of the sheet may or may not be in the direction of the earth's local magnetizing force (= T/μ_0 where T is the total flux density). The magnetic field of the sheet can be calculated by elementary integrations starting from the anomaly of a line pole, p 18 (*164*).

In the notation of Fig. 7a, the z-axis is vertically downwards and the y-axis parallel to the length of the sheet. Then the anomaly $\Delta A(l, 0, n)$ Wb/m² in any arbitrary direction, whose directions cosines are $(l, 0, n)$ at a point P on the line shown, is given by

$$\Delta A(l,0,n,) = \left(\frac{\mu_0}{4\pi}\right) 2M' \sin\theta \; [\{n\cos(\theta - i') - l\sin(\theta - i')\} (\alpha_1 - \alpha_2) -$$

$$\{l\cos(\theta - i') + n\sin(\theta - i')\} \ln (r_1/r_2)\}] \qquad (2.12)$$

Here $\mu_0 = 4\pi \times 10^{-7}$ henry/m, $M' = M(1 - \cos^2 \delta \cos^2 i)^{\frac{1}{2}}$ (amp/m) is the component of M in the cross-section of the sheet, while i is the inclination of M with the horizontal ($-90° < i < +90°$, positive downwards), δ is the angle ('declination') between the strike and the vertical plane through M, and $i' = \tan^{-1} (\tan i/\sin \delta)$ is the inclination of M'. x is positive in the direction from which θ is measured ($0° < \theta < 180°$).

In case M has the same direction as F (that is as T), we can put $\delta = \beta$, $i = I$, and $i' = \tan^{-1} (\tan I/\sin \beta) = I'$ (say) where β is the angle between the strike and the magnetic north.

If in (2.12) $l = 1$, $n = 0$ then $\Delta A = \Delta H$, while if $l = 0$, $n = 1$ then $\Delta A = \Delta Z$. The total field anomaly ΔT in the sense of $|T - T_0|$ (where T_0 is the normal flux density) is given by $(\Delta H^2 + \Delta Z^2)^{\frac{1}{2}}$. The anomaly actually measured by total field instruments is not $|T - T_0|$ but $|T| - |T_0| (\ll |T - T_0|)$ but this is impossible to calculate in a reasonably simple manner.

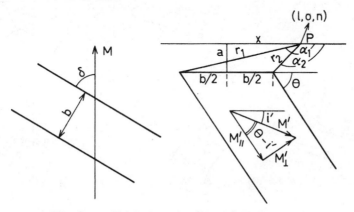

Fig. 7a.   *Thick sheet in plan and section*

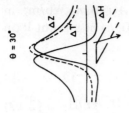
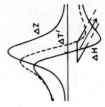
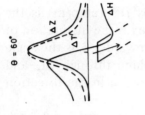
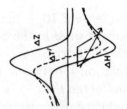
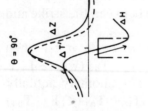
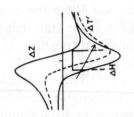
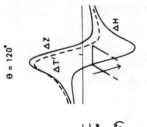
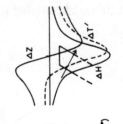
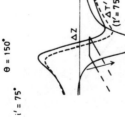
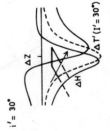
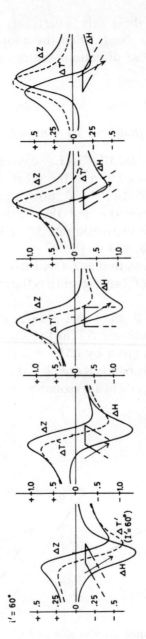

Fig. 7b. ΔZ, ΔH, $\Delta T'$ *across a thick sheet* ($b = 2a$)

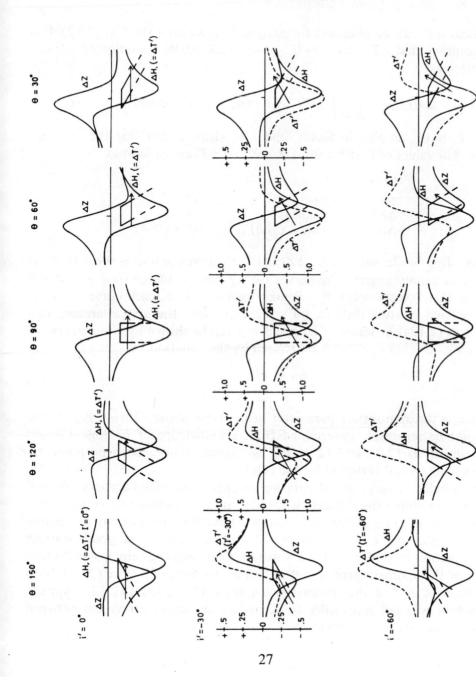

Fig. 7c. ΔZ, ΔH, $\Delta T'$ across a thick sheet ($b = 2a$)

27

A field $\Delta T'$ can be obtained by putting $l = \cos I'$, $n = \sin I'$ in (2.12). It is easy to prove that $\Delta T = \Delta T'/\cos \{I' - \tan^{-1}(\Delta Z/\Delta H)\}$. Equation (2.12) can be written as

$$\Delta A(l,0,n) = \left(\frac{\mu_0}{4\pi}\right) 2M'C\sin \theta \ \{(\alpha_1 - \alpha_2) - k\ln(r_1/r_2)\} \qquad (2.12a)$$

where C is an 'amplitude factor' and k a 'shape factor' for the anomaly profile. The values of C and k for ΔH, ΔZ and $\Delta T'$ are as follows:

Field	C	k
ΔH	$-\sin(\theta - i')$	$-\cot(\theta - i')$
ΔZ	$\cos(\theta - i')$	$\tan(\theta - i')$
$\Delta T'$	$\sin(I' + i' - \theta)$	$\cot(I' + i' - \theta)$

Figs. 7b and 7c show ΔH, ΔZ and $\Delta T'$ curves across a sheet ($b = 2a$) dipping at various angles. The curves for $\Delta T'$ are calculated for $i' = I'$. It will be seen that, except when $\theta - i'$ takes certain values, each curve possesses two extrema (mutual distance $= E$). By expressing the α's (as arctangents) and r's in (2.12a) in terms of x, a and b it can be shown by straightforward differentiation that E, a and b are related by the equation

$$a = \frac{\sqrt{E^2 - b^2}}{2} \frac{k}{\sqrt{1 + k^2}} \qquad (2.13)$$

where k is the appropriate parameter in the table above. Obviously a, b and k cannot be separately determined from a knowledge of E alone. Certain rules similar to (2.8) and (2.9) for estimating a have been proposed by various workers and reviewed by Åm (165).

The simplest manner of interpreting magnetic anomalies across geological features resembling thick sheets is to estimate a trial value of k (for instance by assuming M to be in the direction of F) and by assuming any reasonable θ. For fairly thick sheets ($b \geqslant \sim 2a$) it is usually a good starting approximation to put in (2.13) $b =$ distance between the principal inflection points of the anomaly profile and obtain an estimate of a from (2.13). With the values of a, b, k thus estimated ΔA is calculated from (2.12a) and the parameters adjusted repeatedly to secure a satisfactory agreement between observed and calculated anomalies.

The value of k in the final solution can be used to find θ if this was not known previously. For ΔZ, for instance, $\theta = i' + \tan^{-1}(k)$ provided demagnetization effects transverse to the sheet are negligible (see further below).

The bracketed part on the right-hand side of (2.12) or (2.12a) consists of two terms, one of which $(\alpha_1 - \alpha_2)$ is symmetric in x and the other $(k\ln(r_1/r_2))$ is antisymmetric. This interesting fact has been used (e.g. *166*a, *167*) to construct families of convenient curves for rapid interpretation.

Equations (2.12) and (2.12a) are valid for sheets extending to infinity in either direction perpendicular to the plane of Fig. 7a. The effect of finite length may be estimated from the formula on p. 18. The effect of finite depth extent (a prismatic body) can be easily taken into account by subtracting from the right hand sides of (2.12) and (2.12a) the corresponding expressions for a sheet of identical width and dip but with its upper surface at a depth equal to the depth of the lower face of the prism. The main effect of a finite depth extent is an enhancement of the side extrema, especially on the down-dip side (cf. Figs. 6d, 7d). In Fig. 7d the deeper hole was drilled specifically to verify the inference about the depth extent. The agreement between the interpretation and drilling results is excellent.

Effect of demagnetization.
The value of k obtained in the final adjustment yields $\theta - i'$, the angle between the dip and the inclination of M'. If M' has a remanent as well as induced component, i' is unknown and hence also θ. However, even when there is no remanence, M' may deviate from the known direction I'. This will be understood from Fig. 7e. If the sheet is of infinite strike length the magnetization in its plane will not be $\kappa F' \cos(\theta - I')$ but $\{\kappa/(1 + N_{/\!/}\kappa)\}\, F' \cos (\theta - I')$ where $N_{/\!/}$ is the demagnetization factor. Similarly the transverse magnetization will be $\{\kappa/(1 + N_{\perp}\kappa)\}\, F' \sin (\theta - I')$. For infinite depth extent $N_{/\!/} = 0$ and $N_{\perp} = 1$ (SI) so that $\tan (\theta - i') = \tan (\theta - I')/(1 + \kappa)$. For ΔZ, $k = \tan (\theta - i')$ (p. 28) and

$$\theta = I' + \tan^{-1}\{(1 + \kappa)k\} \qquad (2.14)$$

(If unrationalized emcgs units are used, κ in (2.14) is to be replaced by $4\pi\kappa$.)

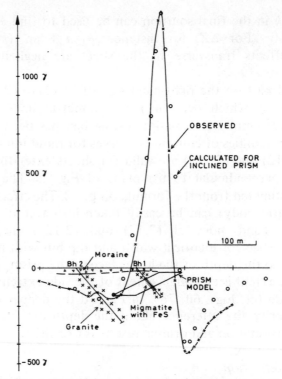

Fig. 7d. *Observed and calculated ΔZ across an inclined prism-like body (k = −1.564)*

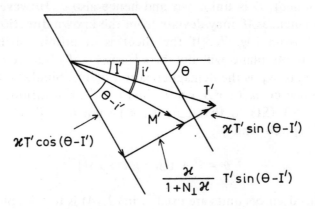

Fig. 7e. *Effect of demagnetization*

For a prism of finite depth extent

$$\theta = I' + \tan^{-1} \left\{ \frac{k(1 + N_\perp \kappa)}{1 + N_{//}\kappa} \right\} \qquad (2.14a)$$

Tables of N (in rationalized system) have been published by Sharma (*168*).

It should be observed that the demagnetization effect tends to deflect M' into the plane of the sheet. The neglect of demagnetization can lead to grossly erroneous conclusions about θ if κ is appreciable.

The interpretation of magnetic anomalies over *thin* sheets can be carried out from formulas obtained by letting $b \to 0$ in (2.12) and (2.12a) and follows essentially the same lines as above. Equation (2.14) is not altered.

Sloping step.
It is interesting to note that the anomalies over this feature are again given by (2.12) and (2.12a) but with the symbols as in Fig. 7f. Analogously to (2.13) it can be shown that in this case

$$a = \left[\frac{E^2 \sin \theta + b^2 (P \cot \theta + \cot^2 \theta - 1)}{P^2 + 4} \right]^{\frac{1}{2}} \qquad (2.15)$$

where $P = 2\,(l\sin i' + n\cos i')/(n\sin i' - l\cos i')$.

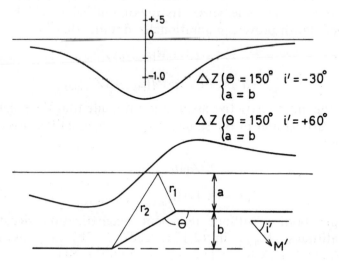

Fig. 7f. *Sloping step*

Unfortunately, although θ can be estimated from the value of k that gives the best fit with the observed data, it is not possible to estimate b by any simple rule, contrary to the case for the thick sheet, so that the adjustments in interpretation are much more difficult to make for the sloping step.

2.10 The Smith rules

The depth to an anomalous structure is a magnitude of prime importance. Some rules for determining it from magnetic observations were given in Section 8 but they presuppose a certain form for the body. A general rule for depth determination does not exist. However, Smith has shown (*8, 12*) that the maximum depth a at which the upper surface of a body causing a magnetic anomaly may lie, can be determined without making any assumption about the shape of the body.

Let $|\Delta Z'|_{max}$ and $|\Delta Z''|_{max}$ denote the absolute maximum values of the first and second horizontal gradients

$$\left(\frac{d}{dx} \Delta Z \text{ and } \frac{d^2}{dx^2} \Delta Z \right)$$

of the vertical field along a measured profile. The magnetization of the body may vary in magnitude but will be assumed to be everywhere parallel though not necessarily in the same sense. Its maximum value will be denoted by $|M|_{max}$. Then Smith proves, in unrationalized units, that

$$a \leqslant \sqrt{(3)} \pi |M|_{max} / |\Delta Z'|_{max} \tag{2.16a}$$

$$a^2 \leqslant 6\sqrt{(2)} \pi |M|_{max} / |\Delta Z''|_{max} \tag{2.16b}$$

If however, the more restrictive assumption is made that the magnetization is everywhere vertical and in the same sense, the inequalities are considerably improved and

$$a \leqslant 2.59 |M|_{max} / |\Delta Z'|_{max} \tag{2.17a}$$

$$a^2 \leqslant 3.14 |M|_{max} / |\Delta Z''|_{max} \tag{2.17b}$$

For isotropic bodies without remanent magnetization situated in high magnetic latitudes, M_{max} in (2.17a) and (2.17b) may be replaced by $\kappa Z / \mu_0 (1 + N\kappa)$ where N is the demagnetizing factor (SI).

For bodies which extend to great distances perpendicular to the measured profile and in which M is everywhere parallel, the Smith rules are that

$$a \leqslant 4M_{max}/|\Delta Z'|_{max} \qquad (2.18a)$$

$$a^2 \leqslant 1.5\pi M_{max}/|\Delta Z''|_{max} \qquad (2.18b)$$

If MKSA units (SI) are used, the right-hand sides of all the formulas in this section must be multiplied by $(\mu_0/4\pi)$.

2.11 Model experiments

Bodies of complex shapes may be simulated in the laboratory by moulding models of appropriate forms. They can be constructed using a uniform mixture of powdered magnetite and plaster of paris or by mixing foundry clay, magnetite, linseed-oil and glycerine (9). There may be placed in a uniform field such as that of a pair of Helmholtz coils and the anomalies over them can be recorded by a suitable device such as the flux-gate magnetometer. Such experiments are used with advantage especially in the interpretation of total field anomalies.

2.12 Some examples of magnetic investigations

The examples below are chosen to illustrate some important points in magnetic interpretation. A general method of attack of any problem is hard to find especially when igneous rocks or chromite and manganese ore bodies, all of which are notorious for the capricious character of their remanent magnetization, are the objects of investigation. Every case of magnetic investigation needs a careful study of the geology and the topography of the area.

(1) *Magnetite ore.*
The map in Fig. 8a shows the results of a recent survey in Central Sweden. The choice of the area was dictated by general geological considerations; the exact location of the magnetic disturbance is the outcome of the geophysical work. The magnetic anomaly shows an approximately E-W strike. From profiles going over points in the immediate vicinity of the anomaly centre the mean $x_{1/2}:y_{1/2}$ (Section 7) was found to be about 0.55 which gave $L/a \approx 3$ and $a/x_{1/2} \approx 1.1$ (Fig. 2). The mean value of $x_{1/2}$ over the central pro-files was 58 m, hence $a \approx 65$ m. The observed and calculated anomalies

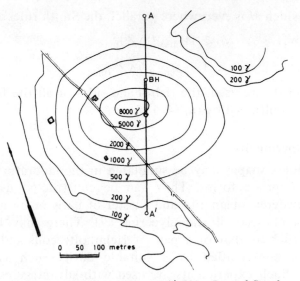

Fig. 8a.   *A magnetic survey (ΔZ) in Central Sweden*

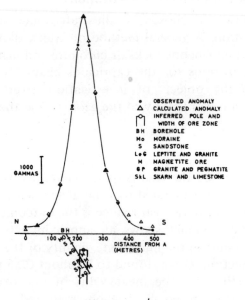

•	OBSERVED ANOMALY
△	CALCULATED ANOMALY
⌐⌐	INFERRED POLE AND
	WIDTH OF ORE ZONE
BH	BOREHOLE
Mo	MORAINE
S	SANDSTONE
L&G	LEPTITE AND GRANITE
M	MAGNETITE ORE
G&P	GRANITE AND PEGMATITE
S&L	SKARN AND LIMESTONE

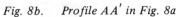

Fig. 8b.   *Profile AA' in Fig. 8a*

using this depth and a magnetic width $(b\kappa) = 86$ also estimated from Fig. 2 are shown in Fig. 8b. A slight effect of the transverse magnetization was also taken into consideration in these calculations. A drill-hole placed as shown encountered rich magnetite ore of total horizontal width of 10 m. This would indicate an average apparent susceptibility of about 8.7 for the ore (cf. Table 1). A hole parallel to the one shown was drilled initially along a line through the anomaly centre but encountered pegmatite at the expected ore depth and, owing to the peculiar disposition of the pegmatite dike, continued to be in it without giving any ore. This illustrates well the uncertainties that lurk in geophysical work even when the anomaly is a 'text book example' and the agreement between observations and calculations is almost as good as might be desired.

(2) *Chromite deposits.*

Magnetic anomalies over two chromite masses (one known before the work) in the Guleman concession area in Turkey (approx. 39°50′ E and 38°30′ N) are shown in Fig. 9 (*10*). According to Yüngül, the susceptibility of the ore masses is, on the average, 2−18 times *smaller* than the surrounding ultrabasic or basic rocks (serpentines, perodotites, norites) so that negative anomalies should be expected over the ores. This is at variance with the observations the positive values of which must, therefore, be attributed to permanent magnetization pointing downwards. Now both Cr_2O_3 and $FeCr_2O_4$ occurring in chromite ores are antiferromagnetic, the former with a weak susceptibility. It is, however, conceivable that the latter compound has ferrimagnetic properties so that spontaneous magnetization of the mass may therefore be possible.

(3) *Sulphide body near Lam (Bavarian Forest).*

The country rock in this area consists mainly of quartzitic shales and evinces a typical layered structure. The dip of the layers is about 70−80° towards the north. The ore occurs in an impregnation zone as veins concordant with the shales. It contains pyrite, chalcopyrite, pyrrhotite and galena carrying values in silver. Some magnetite is also present.

Two magnetic profiles over the ore are shown in Fig. 10 after Zachos (*11*). Maxima in ΔZ and inflection points in ΔH corresponding to each of the two parallel veins are readily evident. The arrows in the lower profile represent total intensity vectors.

In the upper profile are also plotted (1) the susceptibility (κ) (unrationalized units) of the rock samples at different places along an underground gallery leading to the ore and (2) the estimated proportion of the total magnetic constituent in the samples. The susceptibility and ΔZ curves run roughly parallel to each other but the maxima in the former are

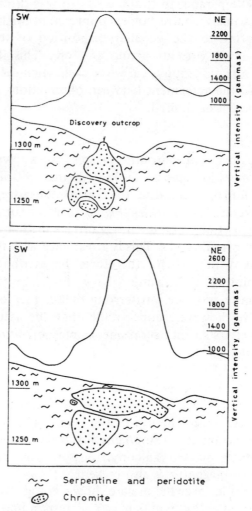

Fig. 9. Δz *profiles across two chromite masses (Ref. 10). After Yüngül (Ref. 10)*

displaced about 10 m to the north. Zachos attributes this difference to the northerly dip of the veins.

The weight ‰ curve departs from the susceptibility curve at several points. This apparent discrepancy has been attributed to variations in the *magnetite:pyrrhotite* proportion in the samples.

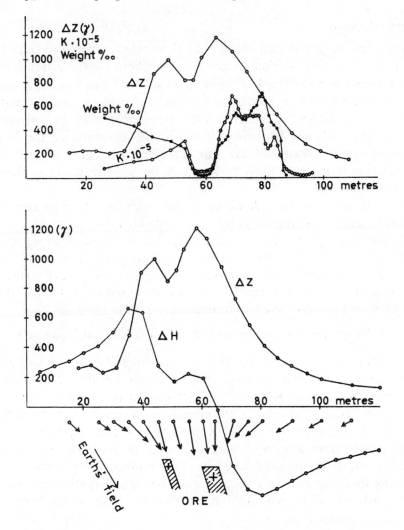

Fig. 10.   *Magnetic profiles across sulphide veins (Ref. 11). After Zachos (Ref. 11)*

3 · Gravitational Methods

3.1 Introduction

Newton's law of gravitation states that the force (in newton) between two point masses m_1, m_2 is equal to $Gm_1 m_2 /r^2$ where r is the distance between the masses and $G = 6.670 \times 10^{-11}$ [$\approx (20/3) \times 10^{-11}$] $m^3 /kg\ s^2$. A unit mass placed in the vicinity of any body will be in a field of force (gravitational field) and experience an acceleration. The force may be calculated by applying Newton's law to infinitesimal volume elements of the body and integrating over the entire volume. The earth has also a gravitational field but in calculating it account must be taken of the centrifugal force due to the rotation of the earth.

The earth may be considered to be an ellipsoid of revolution with an ellipticity (equatorial minus polar radius divided by equatorial radius) = 1/297. The surface of such an ellipsoid of revolution is an equipotential surface. The gradient of the gravitational potential, that is, the force of gravity (g), is, by definition, everywhere perpendicular to the surface which means that it acts in the vertical direction. Its variation with the latitude ϕ, at sea level, can be very closely approximated by the formula

$$g = 9.78049(1 + 0.0052884 \sin^2 \phi - 0.0000059 \sin^2 2\phi)\ m/s^2 \qquad (3.1)$$

This formula was adopted by the International Union of Geodesy and Geophysics in 1930 and is based on an absolute g-value of 981.274 cm/s^2 measured at Potsdam by Kühnen and Furtwängler in 1906.

The most accurate absolute determinations of gravity have been made by means of reversible pendulums of Kater's type. Other methods of absolute determinations have also been suggested, e.g. the free fall of a mass or the determination of the paraboloid of revolution obtained by revolving a vessel containing mercury around a vertical axis. No determinations surpassing the accuracy of reversible pendulum measurements have so far been possible with these methods. In applied geophysics, a knowledge of the absolute gravity is not of immediate interest. We are instead concerned, as in the magnetic methods, with relative measurements. These give the gravity difference Δg between an observation point and a base point. Appropriate corrections (Section 4) must be applied to the differences measured within any region to reduce them to some standard conditions. The corrected Δg-values, called the anomalies, yield information about the changes of density within the earth as well as about the surfaces that bound regions of differing density. The information is, however, always subject to certain fundamental ambiguities inherent in the theory of the Newtonian potential (Section 8).

Gravity anomalies are expressed in milligals. One milligal is one thousandth of a gal which is the c.g.s. unit of acceleration (1 cm/s^2) named after Galileo. A tenth of a milligal is called gravity unit (g.u.) and is one millionth of the SI unit of acceleration (m/s^2). Since the value of g given by equation (3.1) varies between the relatively narrow limits of 9.78049 and 9.83221 m/s^2 from the equator to the poles, one g.u. is roughly one ten-millionth (10^{-7}) of the normal gravity at any place on the earth. The maximum gravity anomalies (on the surface) due to concealed features such as salt domes, oil bearing structures, ore bodies, undulations of rock strata, etc. seldom exceed a few tens of g.u. and, in fact, are often only a fraction of a few g.u. Away from such maxima, the distortions in the normal gravity field of the earth will be even smaller, say, 1–10 parts in 10^8.

3.2 Gravimeters

It is clear from the above that relative gravity measurements, if they are to have any wide application, must be made with an accuracy better than a few parts in 10^7 and, preferably, with an accuracy approaching 1–5 part in 10^8. This aim is achieved in instruments known as *gravimeters*. A number of ingenious gravimeter designs have been proposed during the last fifty years but fundamentally they fall into only two categories, the stable and the unstable types. To this may be added a third type, namely the dynamic but this has seldom been used for geophysical purposes. The stable gravimeter can be described briefly as a highly sensitive balance. It contains a responsive element, usually a spring carrying a weight, which is displaced from the equilibrium position when the force of gravity changes. The displacements are always extremely small (of the order of a few Ångström units) and must be magnified optically, mechanically or electrically. The unstable gravimeter is designed so that when its sensitive element is displaced due to a change in the gravity, other forces tending to increase the displacement come into play. The gravity change can be measured by the force necessary to return the element to its equilibrium position. Brief descriptions of some gravimeter designs are given below to illustrate the principles. More detailed descriptions of these and other gravimeters may be found elsewhere (*14, 15, 16*).

Stable types

(1) *Askania*. In this instrument (Fig. 11a) a beam carrying a mass at one end is held horizontally by means of a main spring (*S*). A mirror placed on the mass reflects a light beam into a double photoelectric cell. The movement of the mass due to a change in the gravity is indicated by the deflection of a galvanometer through which the differential current from the photoelectric cell is led. The mass is restored to the equilibrium position by varying the tension in the auxiliary spring (*S'*). Calibration can be effected by means of small known weights brought on the beam by tilting the instrument.

(2) *Gulf* (*Hoyt*). This gravimeter utilizes a helical spring formed from a ribbon (Fig. 11b). One end of the spring is rigidly clamped while the free end carries a mass with a mirror. An elongation of a helical ribbon spring is always accompanied by a rotation of the free end. In the Gulf gravimeter the rotation is much greater (and therefore can be read more accurately) than the elongation (or the contraction) of the spring caused by a change in the gravity. The range of the instrument is only about 300 g.u. so that a readjustment of the tension in the helix is necessary if gravity differences larger than this amount are to be measured. The accuracy is of the order of 0.2–0.5 g.u.

(3) *Nörgaard*. This is one of the gravimeters combining a wide range (about 20,000 g.u.) with a relatively high accuracy (about 1 g.u.) A small quartz beam carrying a mirror *A* is supported horizontally from a quartz thread, the torsion in the latter counteracting the force of gravity. The mirror *A* is initially parallel to the fixed mirror *B* as is indicated by the coincidence of two index lines in the field of a telescope. When the beam deflects due to a change in the gravity, coincidence can be achieved again by tilting the entire frame through an angle θ. There are two such positions of the frame, one on each side of the initial position. At coincidence the torsion

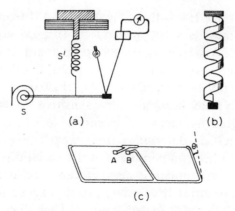

(a)   (b)

(c)

Fig. 11.   Stable gravimeters

moment of the thread must always be the same ($mg_0 l$) so that if g and g_0 are the gravity values at two stations then $g \cos \theta = g_0$.

The instrument can be calibrated by tilting it at small known angles.

Unstable types

(1) *LaCoste-Romberg.* This gravimeter is essentially an adaptation of the long period LaCoste seismograph (*17, 204*) which uses a 'zero-length' spring. Such a spring is wound so that its extension is equal to the distance between the points at which its ends are fastened. Thus, the length defined as the actual physical length minus the extension is zero. The zero length spring S (Fig. 12a) is attached rigidly to the frame at C and balances the mass M at the end of a beam. With the geometry as in the figure, it is easy to show that the net torque on the mass is $(Mg \times \overline{AM} - k \times \overline{AB}^2) \sin \theta$ where k is the spring constant. If $Mg \times \overline{AM} = k \times \overline{AB}^2$ the torque becomes zero, the period infinite and the equilibrium unstable. The instrument is then very sensitive to variations in g.

Readings are taken by restoring M to the original position by raising or lowering C by means of a screw with a calibrated dial. The accuracy is of the order of 0.2 g.u.

(2) *Worden.* The principle of this instrument is very similar to that of the LaCoste-Romberg gravimeter. The mass M (Fig. 12b), is kept in unstable equilibrium by the zero length quartz spring, BC, whose one end is attached to an arm AB inclined at a fixed angle α to AM. Both AM and AB are hinged

at A to a torsion thread. If θ is the deflection of AM from the horizontal the net torque on the mass is easily seen to be $(mg\overline{AM} \cos \theta - \tau (\theta + \theta_0) - k\overline{AC} \times \overline{AB} \cos (\alpha + \epsilon))$ where θ_0 is the permanent torsion in the quartz thread and τ is the torsion constant. By suitable choice of the different constants and the position of C ($\epsilon \approx 0$), the equilibrium can be made unstable and the system becomes very sensitive to variations in g. The equilibrium is restored by means of auxiliary springs arranged as shown in Fig. 12c, one of which determines the range of g measurable by the instrument and the other compensates for the variations in g for a particular setting of (*1*). The instrument is temperature compensated by auxiliary quartz springs and moreover the entire system, except for the reading dials, levels, etc. is kept in a small sealed thermos flask. The total weight of the

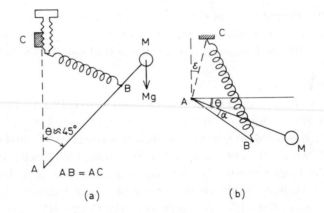

(a)   (b)

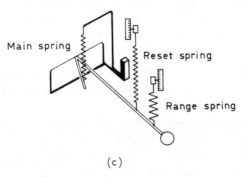

(c)

Fig. 12.   Unstable gravimeters

instrument including the case is about 5 kg but the mass M (made out of fused quartz) weighs only a few milligrams. The accuracy is 0.1–0.2 g.u. and the range of the instrument is wide, namely about 20,000 g.u.

3.3 Field procedure

Gravimeter observations are usually made at the corners of a square grid. The length of the side of the square will depend upon the anticipated dimensions of the features to be located. In oil prospecting the grid side may be of the order of 0.5–1 km or more, while in mineral exploration the stations must often be spaced on a grid with sides no larger than 10–50 m. For other purposes, such as the location of dikes or geological faults, the spacing may be anywhere between these extremes. In large-scale or regional surveys it is also the general practice to establish gravimeter stations along roads.

The geographical positions and the elevations of gravimeter stations must be accurately known in order to reduce the readings to standard reference conditions, as described below. The elevations may be determined by spirit-levelling or barometrically (*18*).

The readings of all gravimeters drift more or less with time, due to elastic creep in the springs. This apparent change in the gravity at a station may be from a few tenths of a g.u. to about 10 g.u. per hour. In order to correct for it, the measurements at a set of stations are repeated after 1–2 h and the differences obtained are plotted against the time between two readings at a station. A 'drift curve' can then be drawn and the corrections read off it. In accurate work it is advisable to determine the drift curve by a least squares adjustment. This is usually straightforward since most gravimeters drift linearly with time. Parabolic or other drift functions are, however, not uncommon.

3.4 Corrections to gravity observations

The gravity difference between two stations is in part due to factors other than the attraction of unknown anomalous masses. These factors and the corrections due to them are as follows.

(1) *Latitude.*

The value of gravity increases with the geographical latitude. By differentiating equation (3.1) we get

$$\frac{dg}{d\phi} = 51.723 \sin 2\phi \quad \text{g.u./rad}$$

If the latitude difference between two stations is small the correction becomes

$$\delta g = 0.081 \sin 2\phi \quad \text{g.u. per 10 m (north-south)} \qquad (3.2)$$

It must be subtracted from or added to the measured gravity difference according as the station is on a higher or lower latitude than the base station. The correction is linear to distances in the N−S direction of the order 1−2 km (about 0.5−1 min of latitude) on either side of the base. If the measurements extend beyond this distance a new base station must be selected and the difference between the normal gravity at it and the first base must be determined by reference to equation (3.1). Extensive tables of normal gravity have been published by the U.S. Coast and Geodetic Survey (1942).

If the north-south distance of a station from the base is known to within 10 m, an accuracy which it is normally not in the least difficult to achieve, the latitude correction will be known to better than one tenth of a g.u.

(2) *Elevation.*
The force of gravity outside the earth varies in inverse proportion to the square of the distance from the earth's centre. If g_1 be its value at the datum level (not necessarily the sea level), then at a height h above it,

$$g = g_1 \frac{R^2}{(R + h)^2} \approx g_1(1 - 2h/R) \qquad (3.3)$$

if powers of h/R higher than the first are neglected. In most practical cases the distance of the datum level from the earth's centre may be taken to be equal to the radius of the earth (6,367 km). The correction for the elevation (the 'free-air correction') is then

$$\delta g = \frac{2g_1}{R} h = 3.086h \quad \text{g.u. (} h \text{ in metres)} \qquad (3.4)$$

The amount 3.086 g.u. per metre of elevation must be added to a measured gravity difference if the station lies above the datum level and subtracted if it lies below it.

If an accuracy of 0.1 g.u. is aimed at in relative gravity measurements, elevation differences from the datum level must be known to better than 4 cm.

(3) *Material between station levels.*

It will be realized by reference to Fig. 13a that while the gravity at B will be less than that at A by an amount $2g_1 h/R$ (free-air correction), it will be greater by an amount $\delta g = 2\pi G\rho h$. This is the additional attraction exerted on a unit mass by the slab of rock material of density ρ between the levels of A and B. The correction

$$\delta g = 0.4191\rho \quad \text{g.u. per metre (of elevation)} \qquad (3.5)$$

is called the Bouguer correction. It must be subtracted from the measured gravity difference if a station lies above the level of the base station and added if it lies below. Throughout this chapter ρ is in g/cm^3 (see page 49).

When measurements are made below the earth's surface the slab of material between A and B (Fig. 13b) exerts an attraction on a unit mass placed at A as well as B. These attractions being in opposite directions the difference of gravity between A and B due to the slab is $4\pi G\rho d$ and the Bouguer correction is doubled (0.8382ρ g.u. per metre).

(4) *Topography.*

At a point such as C, a topographic irregularity (hill, knoll, slope, etc.) will exert an attraction directly proportional to its density. The vertical component ($T\rho$) of this attraction will be directed upwards and reduce the gravity at C. A term of this magnitude must therefore be added to the measured value of gravity at C. A valley such as that near C' is a negative mass and the vertical component of its attraction will also be directed upwards leading again to an *additive* topographic correction.

The topographic correction is calculated by dividing the area around a station in compartments bounded by concentric rings and their radii drawn at suitable angular intervals (ϕ). The mean elevation (z) in each compartment is determined from a topographic map, without regard to sign, that is by treating a hill as well as a valley as positive height differences from the station level. The correction due to the attraction of the material in such a compartment is

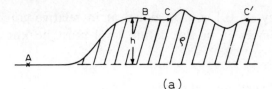

(a)

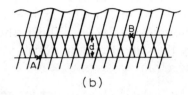

(b)

Fig. 13.   *Rockmass between station and datum plane*

$$\delta g = T\rho = G\rho\phi [r_2 - r_1 + \sqrt{(r_1^2 + z^2)} - \sqrt{(r_2^2 + z^2)}] \qquad (3.6)$$

where r_1, r_2 are the radii of the inner and outer rings bounding the compartment.

Accurate tables of the bracketed expression in (3.6) have been published by Haalck (*15*) and Hammer (*19*).

(5) *Tides.*

The attractions of the sun and the moon may change the gravity at a station cyclically with an amplitude of as much as 3 g.u. during the course of a day. The correction cannot be calculated in any simple way and recourse must therefore be had to tables regularly published in advance for each year (*20, 21*). The drift correction to gravimeter readings includes, in part, the tidal correction.

3.5 The Bouguer anomaly

It will be seen now that the corrected gravity difference between a station and a base is (provided ρ is in g/cm^3)

$$\Delta g_{corr} = \Delta g_{obs} + 3.086 \, h - 0.4191 \, h\rho + T\rho \text{ g.u.} \qquad (3.7)$$

where h is positive if the station is above the base and negative if it is below. The latitude and tidal corrections are included in the term Δg_{obs}. The density of the topographic irregularity in Fig. 13a is assumed to be the same as that of the infinite slab, namely ρ. This assumption may not be always justified. The numerical coefficient of the third term on the right-hand side must be doubled if measurements are made underground. In such measurements T may be negative if, for instance, tunnels are situated *above* the level of a station.

If Δg is to be expressed in mgal the coefficients of h and $h\rho$ in (3.7) must be divided by 10.

The difference Δg_{corr} is called the (relative) Bouguer anomaly. Its variations are to be attributed to the variations of density below the datum level implied in equation (3.7) provided the Bouguer correction also includes the attraction of any inhomogeneities in the slab of material between the surface and the datum level. The reader should also guard himself against the loose expression that the Bouguer anomaly is the 'gravity reduced to the datum level'. It should be realized that the gravity measured on the datum level, at a point B' vertically below B in Fig. 13a, will *not* differ from that at B by Δg_{corr}, because the effect of any inhomogeneities below the datum level is not the same at B as at B'.

3.6 Density determinations

Strictly speaking, any value may be chosen for ρ for the reductions in equation (3.7). However, it is clearly desirable to eliminate the effect of surface features as far as possible by using the true (average) value for their density.

The density may be estimated by laboratory measurements on samples of the rocks exposed within the area of interest. But such estimates suffer from the fact that the samples may be weathered or in other ways unrepresentative. Furthermore, the density of the rocks at depth may be different from that of the surface samples, owing to a variable water content and, in the case of 'loose' rocks like clays, marls, moraine, etc., owing to significant compaction even at moderate depths. Therefore various 'field' methods have been suggested for the determination of ρ. In Nettleton's method (22), the Bouguer anomalies at the stations on a line of measurement are calculated assuming different values of ρ (2.0, 2.1, . . . 3.0, 3.1 . . . g/cm^3). The anomalies that show the least correlation with the topography are adopted as the true anomalies, the corresponding ρ being adopted as the true average density of the surface rocks. Nettleton's original method is graphical, but Jung (*23*) has pointed out that it can be translated into exact mathematical language by putting the correlation coefficient between Δg_{corr} and h equal to zero. We get then

$$\rho = \rho_0 + \frac{\Sigma\ (\Delta g_{corr} - \overline{\Delta g}_{corr})(h - \bar{h})}{0.4191\ \Sigma\ (h - \bar{h})(h - \bar{h} + T - \bar{T})} \tag{3.8}$$

where ρ_0 is an appropriate assumed value of ρ to which a 'correction' term must be added to obtain the true ρ.

In a method proposed by Parasnis (*24*), $\Delta g_{obs} + 3.086h$ in equation (3.7) is plotted against $(-0.4191h + T)$ and the slope of the straight line (determined by least squares) is adopted as the true ρ. This is equivalent to assuming Δg_{corr} (the Bouguer anomaly) to be a random error. In this case

$$\rho = \rho_0 + \frac{\Sigma\ (\Delta g_{corr} - \overline{\Delta g}_{corr})(h - \bar{h} + T - \bar{T})}{0.4191\ \Sigma\ (h - \bar{h} + T - \bar{T})^2} \tag{3.9}$$

This is essentially a generalization of a method due to Siegert (*25*) in which the topographic correction is neglected. Jung has pointed out that while (3.8) and (3.9) reduce to identical forms if $T = 0$, the two values differ but little (a few parts in 1,000) in an actual case even if $T \neq 0$. Equation (3.9) and the associated equation for determining the standard deviation of ρ are considerably simpler from the computational point of view than (3.8).

Legge (*26*) has described a method (neglecting T) in which Δg_{corr}, instead of being treated as a random error, is developed as a power series in the distance of a station from the base.

The above methods deal with stations along a 'line'. The density obtained from (3.8) or (3.9) is then strictly speaking valid only for these stations and the intermediate points. Legge, in the above paper, and Jung (*27*) have shown that the least squares method leading to (3.9) can be generalized to apply to a set of measurements over an area also.

3.7 Interpretation

The interpretation of gravity anomalies in terms of subsurface mass distributions generally follows a pattern similar to the interpretation of magnetic anomalies (Chapter 2). The mass distribution is assumed to correspond to some plausible simple structure and the parameters of the structure are adjusted until its calculated anomaly at all points agrees satisfactorily with the observed anomaly. For this purpose we need to know the gravity anomalies produced by a variety of type-structures (Fig. 14). Some formulas for such anomalies will be given below without proof. Their derivation is generally elementary. We shall hereafter denote the Bouguer anomaly (Δg_{corr}) by Δg. The anomalies in gravity arise from relatively small differences (δ) between rock formations and the interpretation is naturally very sensitive to the available density values. It is, therefore, important to obtain reliable density estimates of the rocks in an area, preferably by a combination of the laboratory and field methods of the previous section. The densities of some commonly encountered rocks and minerals will be found in Table 5.

The SI unit of density is kg/m^3 but on account of the extremely widespread use and the convenient size of the unit g/cm^3 for geophysical purposes, this unit will be retained in the present book. All the formulas of this chapter imply that ρ is expressed in g/cm^3. The unit g/cm^3 is the same as the SI unit megagramme/m^3 which is appropriate for large volumes of rocks.

TABLE 5

Densities (g/cm^3)

Oil	0.90	Granite	2.50–2.70
Water	1.00	Anhydrite	2.96
Sand, wet	1.95–2.05	Diabase	2.50–3.20
,, dry	1.40–1.65	Basalt	2.70–3.30
Coal	1.20–1.50	Gabbro	2.70–3.50
English Chalk	1.94	Zinc blende	4.0
Sandstone	1.80–2.70	Chalcopyrite	4.2
Rock salt	2.10–2.40	Chromite	4.5–4.8
Keuper marl	2.23–2.60	Pyrrhotite	4.6
Limestone		Pyrite	5.0
(compact)	2.60–2.70	Haematite	5.1
Quartzite	2.60–2.70	Magnetite	5.2
Gneiss	2.70	Galena	7.5

(1) *Sphere.*

With the notation of Fig. 14a,

$$\Delta g = \tfrac{4}{3}\pi R^3 \, G\delta \, \frac{\cos\theta}{r^2}$$

$$= \Delta g_{max}/(1 + x_a^2)^{3/2} \qquad\qquad (3.10a)$$

where $\Delta g_{max} = \Delta g$ at $x = 0$ and $x_a = x/a$. It should be noted that even if the cause of an anomaly is *known* to be a spherical body, its radius and density difference from the surrounding rock cannot be separately determined with the knowledge of the gravity field alone. This is because all spheres with the same value for the product $R^3\delta$ ($R \leqslant a$) and having the same centre produce identical gravity fields at the surface.

As an example of an approximately spherical body we may take a boulder. A boulder of 1 m radius is buried in an otherwise homogeneous moraine with which it has a density contrast $\delta = 1.0 \text{ g/cm}^3$. If it just touches the surface of the ground, the gravity anomaly directly above its centre will be 0.28 g.u.

(2) *Horizontal cylinder.*

We take Fig. 14a to represent an infinitely long cylinder striking at right angles to the plane of the paper. Along a line on the ground, perpendicular to the cylinder,

$$\Delta g = \frac{2\pi G R^2 \, \delta a}{r^2}$$

$$= \frac{\Delta g_{max}}{(1 + x_a^2)} \qquad\qquad (3.10b)$$

where

$$\Delta g_{max} = 2\pi G R^2 \, \delta/a.$$

The attraction of an infinitely long cylinder at points outside it is the same as that of an infinitely long rod with the same mass per unit length ($\pi R^2 \delta$). Note that Δg_{max} decreases as the first power of the depth to the cylinder axis.

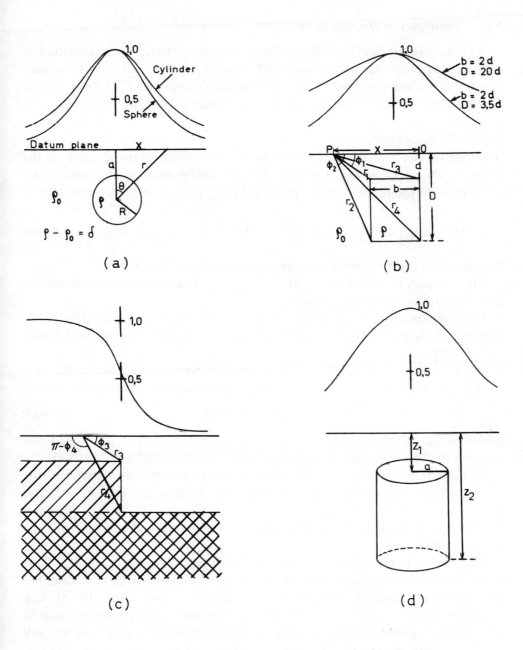

Fig. 14.   *Gravity anomalies across some bodies of simple*
geometrical shape

51

As an example we may note the following. A long, horizontal underground tunnel of circular cross-section (radius 1m), driven in a rock with density 2.7 g/cm³ will produce a maximum decrease of 0.113 g.u. in the gravity at the surface if the axis of the tunnel is at a depth of 10 m.

(3) *Rectangular prism.*

The cross-section of an infinitely long prism striking perpendicular to the plane of the paper is shown in Fig. 14b. The angles made by the lines r_1, r_2, . . ., etc. with the horizontal are denoted by ϕ_1, ϕ_2, . . ., etc. We have then

$$\Delta g = 2G\delta \left[x \ln\frac{r_1 r_4}{r_2 r_3} + b \ln\frac{r_2}{r_1} + D(\phi_2 - \phi_4) - d(\phi_1 - \phi_3) \right] \quad (3.10c)$$

This formula can be derived in a very simple manner (see p. 72).

The rectangular prism is one of the versatile models which can be used to approximate many geological features.

If we imagine the lower boundary face to be extended indefinitely in the horizontal plane, the figure would represent a buried ridge or an undulation of the bedrock.

If the vertical faces be moved in opposite directions to infinity, $r_1 \rightarrow r_2$, $r_3 \rightarrow r_4$, ϕ_1 and $\phi_2 \rightarrow \pi$, ϕ_3 and $\phi_4 \rightarrow 0$ so that equation (3.10c) reduces to

$$\Delta g = 2\pi G\delta (D - d) \quad (3.10d)$$

which is the anomaly of an *infinite slab* of thickness $(D - d)$. It depends only on the thickness and not on the depth to the top surface of the slab.

If the lower surface is moved downwards to a very large distance, the figure would correspond to a vein or a dike. In this case $r_2 \rightarrow r_4 \approx D$ and $D(\phi_2 - \phi_4) \rightarrow b$ in formula (3.10c). The anomaly of a dike increases indefinitely with its depth extent since $r_2 \rightarrow \infty$ in the second term but the increase is very slow.

The step structure shown in Fig. 14c is a special case in which one of the vertical faces of the prism is moved to infinity. The prism is then represented by the singly-hatched portion of the diagram. The anomaly of this configuration ($r_1 \rightarrow r_2$, ϕ_2 and $\phi_1 \rightarrow \pi$) is the same whether the density of the underlying material (the doubly-hatched portion) is ρ or ρ_0 or has any other uniform magnitude. In the first case, (ρ), we have either a buried erosional *escarpment* or alternatively a vertical *geologic fault* in which the rock strata on one side of a vertical plane have been thrust downwards or upwards with respect to those on the other. The 'throw' of the fault is $(D - d)$. The other two cases resemble *flat-lying features* (laccoliths, ore-beds, etc.) with very large dimensions perpendicular to the plane of the figure.

The examples above illustrate the anomalies of a few typical infinitely long bodies. A very full discussion of these as well as finite prismatic bodies has been given by Lindblad and Malmqvist (*28*).

(4) *Vertical cylinder.*
The anomaly of a vertical cylinder (Fig. 14d) at a point P_0 on its axis is given by

$$\Delta g = 2\pi G\delta\,[z_2 - z_1 + \sqrt{(z_1^2 + a^2)} - \sqrt{(z_2^2 + a^2)}] \qquad (3.10e)$$

If $z_2 \to \infty$,

$$\Delta g = 2\pi G\delta\,[\sqrt{(z_1^2 + a^2)} - z_1] \qquad (3.10f)$$

and if, in addition, $z_1 \to 0$

$$\Delta g = 2\pi G\delta a \qquad (3.10g)$$

These equations show that the attraction of a vertical cylinder remains finite as its depth extent increases, in contrast to the attraction of an infinitely deep-going dike.

The attraction of the cylinder at a point P off the axis is often estimated as follows. The attraction of a cylinder of thickness $(z_2 - z_1)$ and radius $(x - a)$ having the vertical line through P as the axis, is subtracted from the attraction of a coaxial cylinder of radius $(x + a)$. The result (valid for $x \geqslant a$) is multiplied by

$$\frac{\pi a^2}{4xa} = \frac{\pi a}{4x}$$

which is the ratio of the area of the plane face of the cylinder in Fig. 14d to the area of the annulus between the two cylinders with the axis through P. For $x < a$, the attraction is estimated by a somewhat more complicated but essentially similar procedure. The method is approximate, however, since the underlying assumption that the anomaly of a cylinder is proportional to the area of its plane face is valid only if the horizontal distance of P is large. The exact expressions for the attraction at P can be derived by using elliptic integrals and are complicated (*29*).

(5) *Bodies of arbitrary shape.*

A surprisingly large number of geological structures can be adequately represented by the above regular shapes and their combinations. Formulas for the gravity anomaly of other regular shapes involving dips or sloping faces have also been derived (*14*). Often, however, it is required to calculate the attraction of irregularly shaped bodies. For this purpose a number of graphical methods are used (*28, 30*). One of them, due to Lindbland and Malmqvist, is illustrated by Fig. 15. The graticule in the figure is so constructed that each field, representing an element of the cross-section of a 100-m long prism, contributes 0.1 g.u. to the gravity field at P. The density of the prism is assumed to differ by 1 g/cm³ (= 1 Mg/m³) from that of the surrounding rock. The prism strikes perpendicular to the plane of the figure, 50 m on each side. If a closed curve is traced on the diagram, the anomaly at P due to the prism with this curve as the uniform cross-section is found by counting the number of fields within the curve. By shifting the graticule and the curve with respect to each other, the anomaly at any point situated in the vertical plane through the central cross-section of the prism may be obtained.

A variety of analogue instruments have also been constructed from time to time for calculating the fields of irregular bodies (*32, 33*). Nowadays, however, the calculations are often made on digital computers. One commonly employed method for this purpose (*34*) approximates the irregular cross-section of the body by a closed polygon, and the body by a number of such polygons on one another. If the sides of the polygon are chosen to be parallel to the x and z axes, a much quicker, though slightly less versatile method, described on page 73, results. The approximation, however, will be found to be adequate for most practical purposes.

3.8 Limitations on gravity interpretation

The discussion in the last section may be called the 'forward' approach to gravity interpretation. That is, given a mass distribution we determine its gravity field on the earth's surface. The real problem in applied geophysics is,

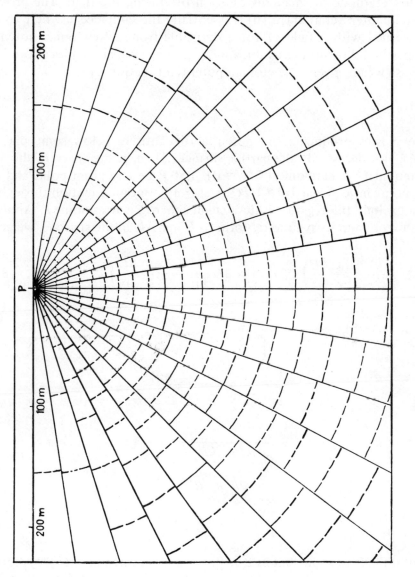

200 m   100 m   100 m   200 m

P

Fig. 15. Sector diagram for calculating gravity (Ref. 28)

however, the inverse one: given a gravity field g over an infinite horizontal plane, to determine the mass distribution producing the field. The problem of magnetic interpretation (p. 16) is essentially the same because here too we are concerned with a field of force derivable from a Newtonian potential. Therefore the following discussion applies to magnetic anomalies also.

Suppose we are given a Newtonian potential function $U(x, y, z)$.

$$g = -\frac{\partial U}{\partial z} \tag{3.11}$$

Let the masses producing the potential lie entirely below some plane A (Fig. 16), not necessarily the earth's surface. Then in the space R above A, the potential U is harmonic ($\nabla^2 U = 0$). Let P be any given point and Q a second point in R. Then $1/r = 1/PQ$ is also a harmonic function throughout R if we exclude the region within a small sphere Σ surrounding P. Applying Green's theorem in its symmetric form to U and $1/r$ in the space R, we get

$$\iint_{R-\Sigma} \left(U \frac{\partial}{\partial z} \frac{1}{r} - \frac{1}{r} \frac{\partial U}{\partial z} \right) dS + \iint_{\Sigma} \left(U \frac{\partial}{\partial n} \frac{1}{r} - \frac{1}{r} \frac{\partial U}{\partial n} \right) dS = 0 \tag{3.12}$$

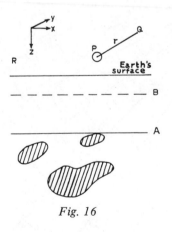

Fig. 16

It is not dificult to show that as $\Sigma \to 0$, the second integral in (3.12) reduces to $4\pi U(P)$ where $U(P)$ is the potential at P due to the masses below A (see, for example, Ref. *35*, p. 219). Hence,

$$U(P) = \frac{1}{4\pi} \iint_A \left(\frac{1}{r} \frac{\partial U}{\partial z} - U \frac{\partial}{\partial z} \frac{1}{r} \right) dS \qquad (3.13)$$

The surface integral reduces simply to that over the boundary A of R. But this equation shows that $U(P)$ is the combined potential of a mass distribution of surface density $-(1/4\pi) (\partial U/\partial z)$ and a double distribution (dipoles) of moment $U/4\pi$, over the plane A. Obviously, if the masses below A were replaced by these two distributions on A, the potential at any point above A will be unaffected. Since U is also harmonic in the subspace above any other arbitrarily chosen plane B, shallower than A, the same reasoning will hold and $U(P)$ will be unaltered if the masses are now replaced by a suitable distribution on B. The inverse problem of gravity interpretation has thus no unique solution.

It might be objected that, in reality, we are measuring only the gravity $g = -(\partial U/\partial z)$ so that U in the second term in (3.13) is unknown. The objection is only apparent for, by setting up a Green's function of the second kind

$$H = \frac{1}{r} + \frac{1}{r'}$$

where r' is the distance of Q from the mirror image of P in the earth's surface, it can be proved that

$$U(x, y, z) = \frac{1}{2\pi} \iint \frac{1}{r} g(x_0, y_0, 0) \, dS \qquad (3.14)$$

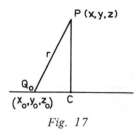

P (x,y,z)

r

Q_0
(x_0, y_0, z_0) C

Fig. 17

In equation (3.14) x, y, z are the coordinates of P and x_0, y_0, 0 of a point on the earth's surface at which the value of gravity is g. Thus, when g is given at all points on the earth's surface the potential can be found not only at a point on the surface ($r = Q_0 C$, Fig. 17) but at any point in the space above. In practice, the integration will be replaced by a summation where it is assumed that g is the mean value of the gravity within a suitably chosen areal element dS.

Equation (3.14) shows that if the masses producing $g(x_0, y, 0)$ are replaced by a thin layer of density $g(x_0, y_0, 0)/2\pi$ the potential in the space above the masses will not be altered in any way. Such a layer is known as Green's equivalent stratum.

Another important consequence of (3.14) is that the derivative of U at x, y, z in *any* direction l can be uniquely determined by the measured distribution of g on the earth's surface. Differentiating (3.14).

$$\frac{\partial U}{\partial l} = \frac{1}{2\pi} \iint g(x_0, y_0, 0) \frac{\partial}{\partial l} \frac{1}{r} \, dS \qquad (3.15)$$

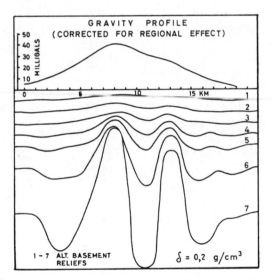

Fig. 18.   *Ambiguity in gravity interpretation (Ref. 36)*

In particular, if $g(x_0, y_0, 0)$ represents the vertical magnetic intensity $Z(x_0, y_0, 0)$ and l the vertical direction,

$$\frac{\partial U}{\partial z} = Z(x, y, z) = \frac{1}{2\pi} \iint -\frac{z}{r^3} Z(x_0, y_0, 0) \, dS \qquad (3.16)$$

Similarly the components of the magnetic intensity in the x and y directions can be determined at any point (on or above the earth's surface) from the distribution of $Z(x_0, y_0, 0)$ alone. Clearly, therefore, horizontal magnetic intensity measurements on the surface of the earth will not yield any additional information if sufficiently dense and precise vertical intensity measurements are available (cf. p. 14).

The inherent indeterminacy of the inverse problem of gravity and magnetic measurements made on the earth's surface will now be appreciated. The interpretation of such measurements in terms of subsurface mass distributions can never be rendered unique unless some external control (e.g. geological information, drill-hole data, etc.) is available. The mere fact that the calculated anomaly of some mass distribution, whose geometrical parameters are appropriately adjusted, agrees everywhere with the measured values, is no guarantee that the distribution occurs in reality, however good the agreement may be.

Further discussions of the ambiguity in gravity and magnetic interpretation will be found elsewhere (*36, 37, 38*). Fig. 18 is an example where a given gravity anomaly has been explained exactly by each of the alternative basement reliefs.

Analytic continuation.
Equation (3.14) is the fundamental integral of upward continuation of potential fields and (3.15) and (3.16) are its special cases. The problem of calculating $U(x, y, z)$ from a knowledge of $\partial U/\partial z(x_0, y_0, 0)$ is a straightforward one of numerical integration. The faithfulness with which U can be reproduced at a higher level, that is farther from the masses, is only limited by the extent of the area on $z = 0$ within which $\partial U/\partial z$ is given, the denseness of the data and the fineness of the mesh (size of $dS = dx_0\, dy_0$) into which the datum plane can be divided for numerical integration.

Transferring the origin to the plane $-z$, (3.16) becomes

$$\frac{\partial U}{\partial z}(x, y, 0) = \frac{z}{2\pi} \int \frac{1}{r^3} \frac{\partial U}{\partial z}(x_0, y_0, z)\, dx_0\, dy_0 \qquad (3.16a)$$

which may be regarded as an integral equation for determining $g(x_0, y_0, z) = \partial U/\partial z(x_0, y_0, z)$ on a deeper level when $g(x, y, 0) = \partial U/\partial z(x, y, 0)$ is given on the surface. This process is called downward analytic continuation. It is a more difficult problem than upward continuation. Its formal solution, however, is easily obtained (Appendix, p. 202) by using the convolution theorem.

The gravity field on a deeper level, calculated analytically starting from the field given on the surface, begins to fluctuate between positive and negative values at some stage. If the given field contains rapid variations of gravity (or magnetic intensity) with distance, indicating shallow sources, the fluctuations start to occur at a comparatively small depth. However, no matter how smooth the original field, the downward continued field will always start to fluctuate at *some* depth. This depth may in certain cases agree with the depth to the top of the masses producing the entire given field but strictly, it represents the maximum depth at which a density distribution of one sign only may lie and explain the original field. However, the reader should note that this information can generally be obtained much more easily by other means (see next section) and the calculation of downward continued fields towards this end does not repay the labour involved (even using digital computers): Further details of the downward continuation problem will be found elsewhere (*45, 54*).

3.9 Depth determinations

If we postulate some regular shape for an anomalous mass, it is usually possible to devise rules which unambiguously determine the depth to its top surface or centre of gravity. As in magnetic interpretation (p. 18), the rules are often based on the distance at which the gravity anomaly or its horizontal gradient falls to a given fraction of the maximum value. Jung (*39, 40*) has derived a number of such rules and also given some geometrical constructions for rapid depth determinations. It is perhaps worth emphasizing that the converse is not true, that is, the geometry of a mass cannot be uniquely determined by assuming a depth to its top surface or centre of gravity. For, it can be shown that if, to any such mass capable of explaining a given gravity field, we add a distribution of the type $A\sin px$ with an arbitrarily great amplitude, the attraction of this extra distribution can be kept less than any quantity, however small, if the wavelength of the distribution is kept sufficiently short.

When no assumptions are made about the anomalous body the shallowest possible mass distribution may evidently lie on the surface of the earth itself, or, if we refer to a volume distribution, it may lie with its top surface just touching the earth's surface at one or more points, unless geological or other data preclude such distributions. On the other hand it can be shown that the top surface of a distribution cannot lie at an arbitrarily *large* depth but that there must be a limiting depth to it. For the rapid estimation of this maximum possible depth (h) from the observed gravity field, Smith (*41, 42*) has obtained a variety of important, rigorous results. Some of them are briefly discussed below. They apply to gravitating bodies whose density contrast with the surrounding rock is either entirely positive or entirely negative. No other restrictions need be placed on the magnitude or variation of the density contrast or on the shape and location of the body. The assumption is thus very light and can be satisfied in a very large number of geological situations.

(1) If Δg_{max} and $\Delta g'_{max}$ are respectively the maximum values of the gravity anomaly and its horizontal gradient, then

$$h \leqslant 0.86 \, \Delta g_{max}/|\Delta g'_{max}| \tag{3.17}$$

The inequality is particularly suitable when Δg_{max} and $\Delta g'_{max}$ occur close to each other.

(2) For all x we have

$$h \leqslant 1.50 \, \Delta g(x)/|\Delta g'(x)| \tag{3.18}$$

This inequality can be used when only part of the whole anomaly is known.

For 'two-dimensional' density distributions such as those in Figs. 14b and 14c the numerical factor may be replaced by 0.65 in the inequality (3.17) and by 1.00 in (3.18).

(3) If $|\delta|_{max}$ is the absolute value of the maximum density contrast and $\Delta g''$ the second horizontal gradient of Δg,

$$h \leqslant 5.40 \, G|\delta|_{max}/|\Delta g''|_{max} \tag{3.19}$$

This inequality may be inverted to find the lower limit of $|\delta|_{max}$ if a plausible value for h can be assigned on other data. If it is assumed that $\delta \geqslant 0$ throughout the body, the inequality can be improved by replacing the symbol $|\delta|_{max}$ by $\frac{1}{2}|\delta|_{max}$.

The above formulas have been generalized by Smith (*43*) in an interesting manner. For instance, suppose that the body lies wholly between the planes $z = h$, $z = l$. Let $\overline{\Delta g}(d)$ be the average value of $\Delta g(x, y)$ around a circle of radius d lying in the plane $z = 0$ and having its centre at $(x, y, 0)$.

$$D = \overline{\Delta g}(d) - \Delta g(x, y, 0) \tag{3.20}$$

Then Smith has shown that

$$|D| \leqslant G|\delta|_{max} d[J(\alpha) - J(\beta)] \tag{3.21}$$

where $\alpha = h/d$, $\beta = l/d$, $J(t)$ is the function plotted in Fig. 19 and $|\delta|_{max}$ is, as before, the maximum density contrast.

The inequality (3.21) can be used in a number of ways depending upon the information already available about the body (see, for example Section 12). If it is assumed that $\delta \geqslant 0$ throughout the body $|\delta|_{max}$ may be replaced by $\frac{1}{2}|\delta|_{max}$.

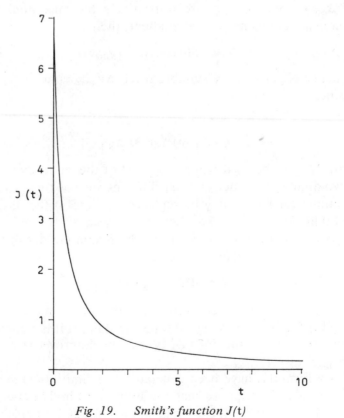

Fig. 19.   *Smith's function J(t)*

3.10 Determination of total mass

It is interesting to note that although gravity measurements alone cannot uniquely determine the distribution of anomalous masses, they do provide a unique estimate of the *total* anomalous mass *(44)*. This is a corollary of Gauss's flux theorem. Let us describe a closed surface bounded by the hemisphere *ABC* (Fig. 20) and its circular intersection *AC* with the earth's surface. The total normal flux across the surface is $4\pi GM$ where M is the anomalous mass enclosed. As *ABC* becomes large, half the flux $(2\pi GM)$ passes across this surface of the hemisphere and the other half across the circle *AC*. From the definition of the normal flux we then obtain,

$$\iint\limits_{AC} \Delta g \ dS = 2\pi GM \tag{3.22}$$

where Δg is the gravity anomaly on *AC*. In practice, *AC* will comprise the area of measurements and the integration will be replaced by a summation. If Δg is expressed in g.u. and the areal element ΔS in square metres, we get from (3.22),

$$M = 2.39 \ \Sigma \ \Delta g \ \Delta S \text{ metric tons} \tag{3.22a}$$

If the body has a density ρ_1 and is embedded in a rock of density ρ_2, its *actual* mass will be

$$2.39 \ \frac{\rho_1}{\rho_1 - \rho_2} \ \Sigma \ \Delta g \ \Delta S \text{ tons} \tag{3.22b}$$

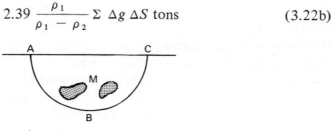

Fig. 20.   *Theorem of Gauss*

It will be seen that the anomalous mass, that is, the difference between the mass of a body and the mass of the country rock occupying an identical volume, can be determined without making any assumptions whatsoever except the physically trivial ones inherent in Gauss's theorem. However, the estimate of the actual mass requires an assumption concerning the *ratio* of ρ_2 and ρ_1.

In equation (3.22), 2π must be replaced by an appropriately smaller solid angle if AC is not sufficiently large compared to the depth of the centre of gravity of M.

3.11 Vertical derivatives of gravity

Let us suppose that we have two point masses at depths a_1 and $a_2 (> a_1)$. Their maximum anomalies Δg_1 and Δg_2 will be in the proportion a_2^2/a_1^2. The maximum values of $\partial \Delta g/\partial z$ will be in the proportion $a_2^3/a_1^3 (> a_2^2/a_1^2)$, those of $\partial^2 \Delta g/\partial z^2$ in the proportion $a_2^4/a_1^4 (> a_2^3/a_1^3)$ and so on. Thus, the successive vertical derivatives of gravity accentuate the relative effect of the shallower point mass. Now, the vertical derivative of gravity of any order, at any point on the earth's surface or above it, can be calculated merely from the knowledge of the distribution of gravity on the earth's surface. This follows from equation (3.15) for we may perform repeated differentiations on $1/r$ with respect to z under the sign of integration and evaluate the resulting double integral numerically. Many different methods of varying precision have been proposed for computing the vertical derivatives from surface data (*45, 46, 47, 48*). We must remember, however, that a knowledge of the vertical derivatives does not in any way reduce the fundamental ambiguity in gravity interpretation. The reason is that the derivatives are directly deducible from the gravity field and therefore do not contain any additional information not inherent in the original data. This fact will not be altered even if the derivatives were directly measured by suitable instruments, alone or in addition to the gravity.

It is now also realized (*55b*) that derivative maps necessarily contain maxima and minima which have little structural significance and are only the result of straighforward algebraic properties.

Another property of the vertical derivatives is as follows. Suppose that we have two equal point masses at the same depth a. The gravity profile will then show two maxima when the masses are separated by a large distance (s) but a single maximum when they are close together. The vertical derivatives will present two principal maxima which also coalesce into one for some value of s. It can be shown (*49*) that whereas the gravity maxima merge into one another when $s = a$, the principal maxima of the second derivative, for example, do not merge until a somewhat smaller value of s, namely $0.639a$, is reached. Thus, if the coalescence of the two maxima is chosen as the only criterion, the vertical derivatives of gravity would seem to have a greater resolving power than the gravity. There are, however, other criteria, e.g. the width of the gravity curve, which yield as much resolving power as the derivatives.

The above discussion concerns idealized cases. In practice, we generally have surface irregularities. Their effect on the vertical derivatives increases rapidly with the order of the derivative so that if, say, the second derivative is measured directly the topographic corrections to it will be considerably more uncertain, than the corresponding corrections to the gravity.

3.12 Illustrations of gravity surveys and interpretation

(1) *West Midlands (England).*
This example is taken from a paper by Cook *et al.* (*50*). The contours of the Bouguer anomalies in the area of measurement are shown in Fig. 21. The general geologic section is as follows. Rocks of Triassic age (Keuper Marl and Sandstones) are underlain by Coal Measures which contain coal seams and rest upon strata of Old Red Sandstone. Beneath the O.R.S. follow the rocks of Silurian and Cambrian age. All the strata are generally flat or have gentle dips but there is considerable faulting. Igneous intrusions occur at some places. The Trias is exposed in the Forest of Arden and in the area between between the Forest of Wyre and South Staffordshire coalfields. In these coalfields and in the Warwickshire coalfield to the east the surface rocks are composed of Coal Measures.

The first step in the interpretation is to estimate the attraction of the surface rocks. Their thickness is fortunately well known over most of the area. However, this estimate cannot be made directly because although the density of the surface rocks may be known, the attraction is proportional to the difference of this density from that of the underlying rocks, which is not known. Cook therefore proceeds by plotting the anomalies against the thicknesses of the surface rocks concerned, say Coal Measures, at the observation stations. It will be appreciated from Section 6 (cf. equation (3.9)) that the points thus obtained will lie approximately on a straight line with a slope 0.0419δ mgal/m ($0.0128\,\delta$ mgal/ft) where δ is the difference of the density of the Coal Measures from that of the rocks below them. The plot for the Warwickshire coalfield (Fig. 22) yields $\delta = 0.23 \pm 0.02$ g/cm^3. Similarly the density difference between the Trias and the underlying rocks

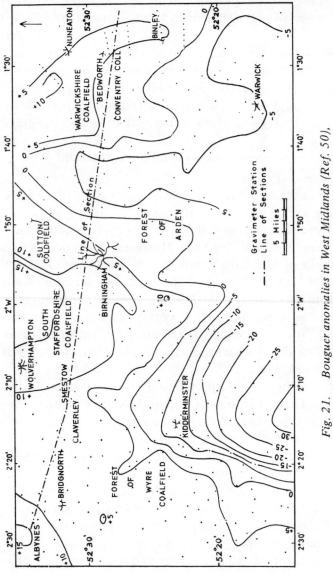

Fig. 21. Bouguer anomalies in West Midlands (Ref. 50), contours in mgal, 1 mgal = 10 g.u.

66

(Coal Measures) is found to be 0.25 ± 0.14 g/cm^3. The Bouguer anomalies then must be corrected for the attraction of the known Trias and Coal Measures, adopting the above density differences.

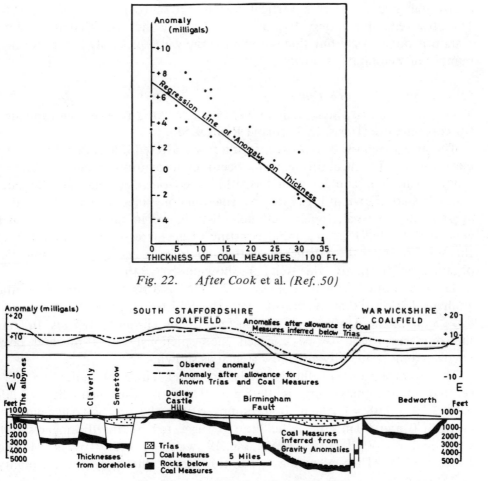

Fig. 22.   *After Cook* et al. *(Ref. 50)*

Fig. 23.   *Section along line in Fig. 21*

Figure 23 shows the residual anomalies along a line from The Albynes to Bedworth. They indicate a large structure under the Trias cover between the Birmingham Fault and the Western Boundary Fault of the Warwickshire coalfield. It seems plausible from the geology that the low anomaly values in this part are caused by a slab of Coal Measures resting on the heavier Palaeozoic rocks. The western boundary of the slab is the known Birmingham Fault. Assuming the eastern one to be a vertical step the thickness of the slab can be estimated from equation (3.10d) to be about 4,000 ft. The maximum horizontal gradient over a vertical step is $2\pi G\delta \ln(D/d) = 48$ g.u./1,000 ft. The observed value is only 10 g.u./1,000 ft which indicates that the eastern margin is not vertical but slopes down (to the west) probably as a result of a number of step faults as shown.

(2) *Camaguey Province, Cuba.*

The measurements discussed here were included in a prospecting campaign for chromite ore (Davis, Jackson and Richter *51*).

The area is relatively flat except for a few scattered hills, mine dumps and excavations. The chromite deposits occur in serpentinized periodotite and dunite near their contact with felspathic and volcanic rocks. The Bouguer anomalies are shown in Fig. 24a. The isoanomalous lines clearly indicate that apart from the local, roughly circular disturbance in the centre, there is a general ('regional') gravity trend consting of an increase in the values from the southwest to the northeast (dotted contours). The residual anomalies obtained on subtracting this trend are shown in Fig. 24b.

The maximum residual anomaly is 2.1 g.u. and its maximum horizontal gradient 1.6 g.u./20 m. Substituting in the Smith inequality (3.17) we get $h \leqslant 23$ m for the depth to the top surface of the chromite deposit. In fact, the ore was found to be exposed inconspicuously in a pit and later uncovered more fully by a shallow trench as shown.

The average value of the anomaly around a circle of radius $d = 20$ m centred at the maximum is 0.98 g.u. Thus, in the inequality (3.20) $|D| = 2.10 - 0.98 = 1.12$ g.u. If we suppose that the ore body lies entirely between 0 and 60 m (the latter limit is suggested by the drilling results), $\alpha = 0$ and $\beta = 60/20 = 3.0$. From Fig. 19 we read off $J(0) = 6.591$ and $J(3.0) = 0.579$. Then assuming $\delta \geqslant 0$ throughout the body the inequality (3.21) becomes

$$\tfrac{1}{2}G\,\delta d \times (6.591 - 0.579) \geqslant 0.112$$

leading to a minimum density difference $\delta = 0.27$ g/cm^3 between the chromite and the surrounding serpentinized peridotite. The latter is

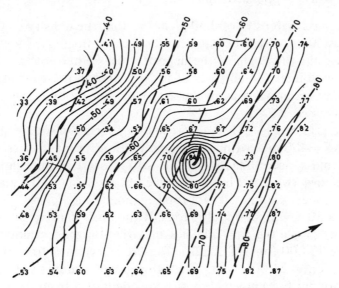

Fig. 24a.   *Bouguer anomalies on a chromite prospect (Ref. 51), contours in mgal,*
1 mgal = 10 g.u.

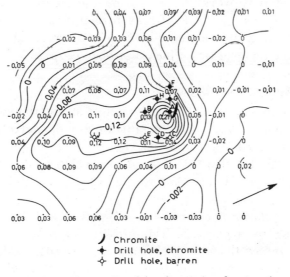

⟩ Chromite
✦ Drill hole, chromite
✧ Drill hole, barren

Fig. 24b.   *Map in Fig. 24a after regional correction*

69

reported to have a density of 2.50 g/cm^3 so that the density of the chromite deposit must everywhere be equal to, or greater than, 2.77 g/cm^3. This, in turn, means that the grade of the deposit must be better than 12.5 per cent chromite (by volume) in all parts of it. Note that in these estimates we have not made any assumption as to the shape of the ore body, its depth, or its location sidewise.

Since the grid in Fig. 24b is a close and regular 20-m square array we may associate $\Delta S = 400$ m^2 with every observation point for estimating the anomalous mass according to (3.22). After inserting interpolated points between the last two rows in Fig. 24b we obtain $\Sigma \ \Delta g = 30.5$ g.u. for the 86 points in the figure. The total anomalous mass is therefore 2.39 x 400 x 30.5 = 29,000 metric tons, If the body causing the anomaly has a density of, say, 4.0 g/cm^3 and lies in serpentinized peridotite its actual mass will be 77,000 tons (cf. 3.22b). The ore ascertained from the drill holes amounts to only 24,000 tons. That the gravity anomaly is not entirely accounted for by the known chromite is evident also from the high anomaly values to the south of the maximum. The remaining mass, of course, need not be chromite.

Regional trends, such as the one in the above example, are often encountered in gravity work. They may be caused by deep-seated masses within the earth's crust, contacts of rock formations with differing density, etc. Apart from the graphical method referred to here, analytical procedures for eliminating the regional trends have also been proposed (*52, 53, 54, 55*a).

Note on the Eötvös Torsion Balance

Before the development of modern gravimeters an instrument known as the Eötvös torsion balance was in considerable use for the detection of the local anomalies in the earth's gravitational field. Its principle, illustrated by Fig. 25, is similar to that of Cavendish's celebrated balance for the determination of *G*. In fact, the latter type of balance could also be used for gravity surveys.

In the Eötvös balance two masses m, m', each about 25 g, are suspended at different levels by a platinum iridium wire. Suppose that an anomalous mass is situated below the instrument and on the reader's side. Its attraction on m will be greater than its attraction on m' so that a couple will act on the balance beam turning it out of the plane of the paper. An equilibrium will be reached when the couple is neutralized by the torsion moment of the suspension. By orientating the instrument in six (strictly speaking only five) different azimuths and observing the deflections in each case, the following quantities, where U is the gravitational potential, can be determined:

$$\frac{\partial^2 U}{\partial x^2} - \frac{\partial^2 U}{\partial y^2}, \frac{\partial^2 U}{\partial x\,\partial y}, \frac{\partial^2 U}{\partial x\,\partial z}, \frac{\partial^2 U}{\partial y\,\partial z}$$

The quantities $\dfrac{\partial^2 U}{\partial x\,\partial z}$ and $\dfrac{\partial^2 U}{\partial y\,\partial z}$ are the gradients of $\dfrac{\partial U}{\partial z}$, that is, of g, in the x and y directions. Adding them vectorially we get the direction and magnitude of the horizontal gradient of gravity at the observation point.

Another quantity that has frequently been used in the interpretation of torsion balance data is:

$$\left[\left(\frac{\partial^2 U}{\partial x^2} - \frac{\partial^2 U}{\partial y^2}\right)^2 + 4\left(\frac{\partial^2 U}{\partial x\,\partial y}\right)^2\right]^{1/2}$$

It can be shown that this quantity is equal to $g(1/R_1 - 1/R_2)$ where R_1 and R_2 are the principal radii of curvature of the equipotential surface at the observation point. $g(1/R_1 - 1/R_2)$, which has the same dimensions as the gradient of gravity, has been called the horizontal directive tendency (H.D.T.) or, by some writers, the curvature difference. To gain some idea of its nature, it may be noted that above a ridge of heavy rock, for example, the H.D.T. vectors will be parallel to the ridge. If the ridge is lighter than the surrounding rock the vectors will be perpendicular to the ridge.

The detailed theory of the Eötvös balance is complicated. For an excellent account of it the reader is referred to Eve and Keys (*56*).

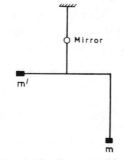

Fig. 25. Eötvös torsion balance

A gradient of 10^{-9} newton/kg-m $= 10^{-3}$ g.u./m($= 10^{-6}$ mgal/cm) is termed an Eötvös unit (E). The northward gravity gradient in latitude 45°N, for instance, is 8.1E (3.2). The torsion balance has a sensitivity of about 1E and certainly ranks as one of the most sensitive physical instruments. Unfortunately, due to the fact that the instrument takes a relatively long time in reaching equilibrium and, due to the necessity of measuring the deflections in five or six azimuths, one complete observation takes several hours. A gravimeter observation on the other hand needs hardly more than 1–5 min. The torsion balance measurements require, moreover, elaborate corrections for the topographic irregularities in the vicinity of a station, including features such as houses, walls, ditches, etc. The modern gravimeters have therefore completely replaced the torsion balance in routine surveys.

Derivation of Formula (3.10c)

Provided the boundary of a closed area is piecewise continuous, an areal integral can be converted into a line integral along the boundary, according to the relation:

$$\iint \frac{\partial u}{\partial z} \, dz \, dx = \oint u \cos (n, z) \, dl$$

where $u(x, z)$ is a continuous, differentiable function, dl denotes an element of the boundary and $\cos(n, z)$ the direction cosine of the normal to it. This theorem due to Gauss will be found to be proved in any standard text on the calculus.

Choosing P as the origin in Fig. 14(b) the vertical component of the gravitational attraction at it due to an infinitesimal element $dx dz$ of the prism (density δ) is given by $2G\delta z dx dz/(x^2 + z^2)$ which is an elementary result (cf. (3.10b)). The total effect of the prism Δg is obtained by integrating this with respect to x and z. But the elemental attraction can be expressed as $G\delta \, dx dz \partial/\partial z \, \{\ln(x^2 + z^2)\}$ so that by putting $u = \ln(x^2 + z^2)$ we get

$$\Delta g = G\delta \oint \ln(x^2 + z^2)\cos(n, z)dl$$

Now, the contour integral reduces simple to one along the two horizontal edges of the prism since along the vertical edges $\cos(n, z) = 0$. Thus

$$\Delta g = G\delta \cos(n, z) \int \ln(x^2 + z^2) dx$$

The integral here is elementary, Evaluating it by parts we get

$$\Delta g = G\delta \cos(n, z) \left[x\ln(x^2 + z^2) - x + z \tan^{-1}(x/z) \right]_{x_1}^{x_2}$$

where x_1, x_2 are the x-coordinates of the corners of the prism. Bearing in mind that $\cos(n, z) = -1$ along the upper face $(z = d)$ and $+1$ along the lower face $(z = D)$ we evelute the bracketed expression, and shifting the origin to 0 in Fig. 14(b), immediately get (3.10c).

The method can be extended to any prism whose cross-section is a right-angled polygon with sides parallel to the x and z axes. Let $1, 2, \dots N$ be the corners (obviously even in number) of such a polygon, numbered anticlockwise, and let x_i, z_i be the coordinates of the corner i with P as the origin. Then the gravity anomaly at P is given by

$$\Delta g = 2G\delta \sum_1^N (-1)^i \{(x_i/2)\ln(x_i^2 + z_i^2) + z_i(\tan^{-1}(x_i/z_i)\}$$

It is very easy to program this equation on a digital computer. Kolbenheyer (*31*) has shown that also the derivatives of the gravitational attraction of the prism can be calculated by similar simple formulas and that the method can be extended to three-dimensional structures without difficulty.

4·Electrical Methods

4.1 Introduction

The electrical properties of the subsurface can be explored either electrically or electromagnetically. The three purely electrical methods to be discussed here are: (*a*) self-potential (SP), (*b*) earth resistivity (ER) and (*c*) induced polarization (IP). The SP method was used by Fox as early as 1830 on the sulphide veins in a Cornish mine (*57*) but the systematic use of the SP and ER methods dates from about 1920. The systematic use of the IP method dates from about 1950 although the associated phenomenon has been known to physicists and chemists for a long time.

4.2 Self-potential

The SP method, as its name implies, is based upon measuring the natural potential differences which generally exist between any two points on the ground. These potentials, partly constant and partly fluctuating, are associated with electric currents in the ground. We shall revert to the fluctuating part in the next chapter. The constant and unidirectional potentials are set up due to electrochemical actions in the surface rocks or in bodies embedded in them.

Ranging normally from a fraction of a millivolt to a few tens of millivolts self-potentials sometimes attain values of the order of a few hundred millivolts and reveal the presence of a relatively strong sub-surface 'battery cell'. Such large potentials are observed as a rule only over sulphide and graphite ore bodies. Over sulphides at any rate they are invariably negative (Fig. 26a). Similarly they are negative on graphite and graphitic shales (*61*) as well as on magnetite, galena and other electronically conducting minerals.

Field procedure.
The measurement of self-potentials is quite easy. Any millivoltmeter with a sufficiently high input impedance may be connected to two electrodes driven some 10–15 cm into the ground and read off. Alternatively a usual

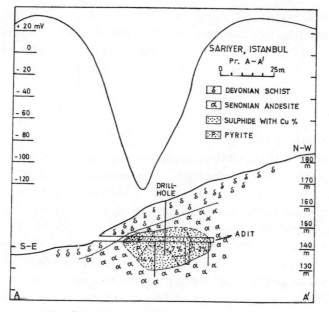

Fig. 26a. *SP profile across a pyrite mass*

potentiometer circuit of the null type (or a pH meter!) can also be employed with advantage. The electrodes must be non-polarizable (e.g. Cu in $CuSO_4$ solution or Pt-calomel in KCl solution); ordinary metal stakes will not generally do so since electrochemical action at their contact with the ground tends to obscure the natural potentials.

Two alternative procedures are in use for SP surveys. The electrodes, say 20 m apart, are advanced together along staked lines or else, one electrode connected to one end of a long cable on a reel is kept at a base point while the other electrode, the reel, and the voltmeter, are carried to different points as the cable is laid off. A new base point is chosen when the cable length is exhausted. The first procedure measures essentially the gradient of the potential.

The *formal* interpretation of SP anomalies is often carried out using formulae closely similar to those for the anomalies of magnetic poles. It must be remembered, however, that SP anomalies are greatly affected by local geological and topographical conditions and these factors must be very carefully considered in SP work. For instance, it is often found that high ground is more negative than low ground, indicating that the electric current tends to flow uphill. An exceptionally high SP value (-1842 mV) associated with alunite (a basic sulphate of Al and K), but to a large extent due to topographic causes, has been reported from a mountain top near Hualgayoc, Peru (*166*b).

Origin of self-potentials.

There has been considered speculation concerning the electrochemical mechanism producing self-potentials. The normal potentials observed over clays, marls and other sediments present no great problem. They can be explained by such well-known phenomena as ionic layers, electrofiltration, pH differences and electro-osmosis. It can be mentioned for instance that water flowing through the sand on a sea-beach sets up electrical potentials due to electrofiltration. However, the large potentials observed over sulphide-ores and graphite are more difficult to explain. Most theories have attributed the sulphide potentials to the oxidation of parts of the ore above the water table, but such an explanation cannot fit graphite which does not normally undergo significant oxidation.

The most detailed theory of the mechanism so far is that proposed by Sato and Mooney (*58*). Since we are concerned with an electronic conductor (the ore) in contact with an ionic conductor (electrolytes in the country rock), there must be an exchange of electrons and ions at their boundary. The directions of current and ion flow implied by a negative centre over a sulphide are shown in Figure 26b. Evidently, electrons are being supplied to the upper end of the ore body, whereas an oxidation of this end will require a liberation of electrons. Sato and Mooney conclude therefore that self-potentials cannot be caused directly by the near-surface oxidation of the ore. They propose instead that there is an oxidation potential difference (*Eh*) between the substances in solution above the water table and those below. The ore body, being a good electronic conductor, merely serves to transport the electrons from the reducing agents at depth to the oxidizing ones at the top, without itself directly participating in the electrochemical action.

Sato and Mooney are able to explain the order of magnitude of the self-potentials observed in the field with the above thoery. The theoretical estimates are, however, definitely on the lower side. This fact and some others

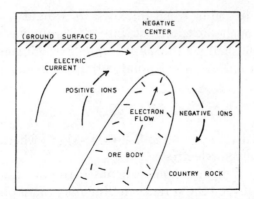

Fig. 26b.   Current and ion flow in the vicinity of an ore body

(e.g. the existence of large potentials even when the ore is totally submerged under the water table) indicate, however, that the problem of origin of sulphide potentials is not yet completely solved.

Finally, it can be mentioned that observations of the electrofiltration or streaming potentials mentioned above have been used in the U.S.S.R. (*169*) to detect water leakage spots on the submerged slopes of earth-dam reservoirs.

4.3 Earth resistivity

In the ER method a direct, commutated or low frequency alternating current is introduced into the ground by means of two electrodes (iron stakes or suitably laid out bare wire) connected to the terminals of a portable source of e.m.f. The resulting potential distribution on the ground, mapped by means of two probes (iron stakes or, preferably, non-polarizable electrodes), is capable of yielding the distribution of electric resistivity below the surface.

The method has been used mainly in the search for water-bearing formations, in stratigraphic correlations in oil fields and in prospecting for conductive ore-bodies.

Potential distribution in a homogeneous earth.
Consider a point electrode on the surface of a homogeneous earth extending to infinity in the downward direction and having a resistivity ρ. Describe a

hemispherical shell of radius r and thickness dr around the electrode. If the current passing through the electrode into the ground is $+I$, the potential difference across the shell will be $dV = - I\rho dr/2\pi r^2$. Integrating, we get for the potential at a distance r from a point current source,

$$V(r) = \frac{I\rho}{2\pi} \frac{1}{r} \tag{4.1}$$

The total potential at any point is $V = V(r) - V(r')$ where r' is the distance from the negative current electrode.

It can be shown that in a homogeneous earth the fraction of the total current confined between the surface and the horizontal plane at a depth z is $(2/\pi) \tan^{-1} (z/L)$ where L is half the distance between the current electrodes.

We see from this that as much as 50 per cent of the total current in a homogeneous earth never penetrates below the depth $z = L$ and as much as 70.6 per cent never below $z = 2L$. The current will evidently penetrate deeper the greater the electrode separation.

The current penetration in a non-homogeneous earth composed of strata with horizontal interfaces has been worked out by Muskat and Evinger (59). Fig. 27 is reproduced from their paper.

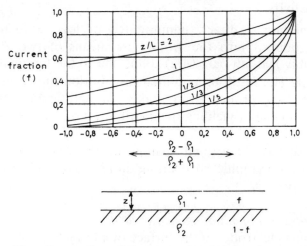

Fig. 27.   *Penetration of electric current in a two layer earth*

Apparent resistivity.

Let C_1, C_2 be the current electrodes, positive and negative respectively, and P_1, P_2 the potential probes. If ΔV is the potential difference between P_1, P_2 it follows from (4.1) that

$$\rho = 2\pi \frac{\Delta V}{IG} \tag{4.2}$$

where

$$G = \frac{1}{C_1 P_1} - \frac{1}{C_2 P_1} - \frac{1}{C_1 P_2} + \frac{1}{C_2 P_2}$$

($\Delta V/I = R$ is of the nature of a resistance). It is advisable to use a null method in measuring ΔV in order to eliminate the influence of the contact resistances at P_1, P_2.

In an actual case, ρ will vary on altering the geometrical arrangement of the four electrodes or on moving them on the ground without altering their geometry. That is, R will not be directly proportional to G as on a homogeneous earth. The value of ρ, obtained on substituting the measured R and the appropriate G in (4.2), is called the *apparent resistivity* (ρ_a). It can be calculated for given electrode arrangements for a number of subsurface resistivity distributions.

The apparent resistivity is a formal, rather artificial concept and should not be construed, in general, as representing the average resistivity of the earth or any other similar magnitude. The artificiality will be evident from the fact that negative values of apparent resistivity are perfectly possible (*60*). For a proper assessment of this quantity one must always bear in mind the configuration with which it has been obtained.

Electrode configurations.

A variety of electrode arrangements are possible (*61, 85*) but we shall restrict ourselves to three very common linear arrays ($C_1 P_1 P_2 C_2$) shown in Fig. 28.

In the *Wenner* configuration the separations between the adjacent electrodes are equal (a) so that (4.2) reduces to

$$\rho_a = 2\pi a R \tag{4.3}$$

In the *Schlumberger* configuration $P_1 P_2 (= 2l) \ll C_1 C_2 (= 2L)$. At the centre of the system, equation (4.2) can be written as,

$$\rho_a = \frac{\pi L^2}{2l} \frac{\Delta V}{I} = \frac{\pi L^2}{I} \frac{dV}{dr} \tag{4.4}$$

where, obviously, dV/dr is the surface gradient of V, that is, the electric field at the centre. If $l \leqslant L/5$, equation (4.4) gives a ρ-value correct to within a couple of per cent.

In the *dipole-dipole* array $P_1 P_2$ are outside $C_1 C_2$, each pair having a constant mutual separation a. If na is the distance between the two innermost electrodes (C_2, P_1) then equation (4.2) gives

$$\rho_a = \pi n(n + 1)(n + 2)a \frac{\Delta V}{I}$$

Measurements are taken by increasing n in steps. If $n \gg 1$ each electrode pair may be treated as an electric dipole. Non-collinear dipole arrangements have also been used, especially in the U.S.S.R. The ρ_a values obtained with them bear simple relations to the value obtained (4.4) when the Schlumberger array is used (*85*).

It may be of interest to note that, in virtue of Helmholtz's reciprocity theorem in the theory of electric circuits, the ratio $\Delta V/I$ and hence ρ_a will be unaltered if the current and potential electrodes (in any configuration) are interchanged. This is true even if the earth is nonhomogeneous.

The properties of the subsurface may be explored by two main procedures often called, by analogy, electric sounding (or drilling) and electric mapping (or trenching).

In electric drilling with the Wenner arrangement the distance a is increased by steps, keeping the midpoint of the configuration fixed. If the Schlumberger configuration is used, only the current electrodes are moved outward symmetrically, keeping $P_1 P_2$ fixed at the centre.

In electric trenching with the Wenner method an electrode array with a fixed a is moved on the earth (usually along staked profiles) and the apparent resistivity is determined for each position. Alternatively, the two potential electrodes having a relatively small, constant separation in the Schlumberger configuration are moved between the fixed current electrodes to different positions. In this case, G in equation (4.2) will vary with the position of the probes.

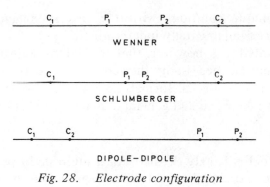

Fig. 28.   Electrode configuration

Since the fraction of the current penetrating to deeper levels increases with the electrode separation, electric drilling primarily provides information about the variation of electric conductivity with the depth. On the other hand, electric trenching measurements are influenced in the first place by lateral variations in the conductivity. They are therefore best suited for detecting local, relatively shallow non-homogeneities in the ground. The two procedures are thus complementary to a large extent.

Inverse problem of ER measurements.
The problem of determining the distribution of the conductivity σ within the ground, when the surface potential due to a current electrode is given amounts (in isotropic media) to integrating the equation

$$\text{div}(\sigma \nabla V) = 0 \qquad (4.5)$$

It was shown by Slichter (*64*) and Langer (*65*) that if σ is a function of depth only the above equation possesses a unique solution. Then the subsurface distribution of σ can be calculated from a knowledge of the surface potential produced by a single point electrode without any further physical or geological data.

In the general case when σ is not a function of z only, Stevenson (*66*) proved that equation (4.5) possesses no unique solution, but that the problem can be made determinate by adding another 'dimension'. Thus, for example, if the surface potential is measured everywhere for *all* positions of a point electrode on some curve on the surface, the distribution of σ can, in principle, be uniquely determined.

It is instructive to contrast the present inverse problem with its theoretical possibilities of unique solutions with the inverse magnetic and gravity problems which are fundamentally indeterminate.

It must be pointed out, however, that the determination of σ solves the *electric* problem but not necessarily the *geological* one, the reason being that given rock types and geological formations are not associated with definite resistivities except in a broad and general manner (see Section 4.8).

4.4 Layered earth

Stevenson's theorem is largely of theoretical interest. In practice one must, as a rule, resort to a trial-and-error technique in which the surface potential due to a current source is calculated for an assumed conductivity distribution and compared with the observations.

One type of conductivity distribution that adequately describes many geological situations is that represented by an earth composed of several horizontal strata. This model is of particular importance in prospecting for ground water by earth resistivity methods.

In the simplest case we have a horizontal layer of thickness h_1 and resistivity ρ_1 overlying a second homogeneous medium of resistivity ρ_2 (Fig. 29). The potential of a point electrode C, through which a current I is passing into such an earth, was first calculated by Hummel (67) by the method of electric images. The potential is given by the sum of (i) the potential of C in a semi-infinite medium of resistivity ρ_1, that is $I\rho_1/2\pi CP$ according to equation (4.1), and (ii) the potentials of fictitious current sources C', C'', C''', ..., etc. of appropriate strengths, where C' is the image of C in the plane $z = h$, C'' is the image of C' in $z = 0$ and so on ad infinitum. We then get for the potential at a surface point P:

$$V(r) = \frac{I\rho}{2\pi}\frac{1}{r}\left[1 + \sum_{n=1}^{\infty}\frac{2k^n r}{(r^2 + 4n^2 h^2)^{1/2}}\right] \qquad (4.6)$$

where $r = CP$ and $k = (\rho_2 - \rho_1)/(\rho_2 + \rho_1)$.

This equation can also be expressed in the closed form

$$V(r) = \frac{I\rho}{2\pi}\frac{1}{r}\left[1 + 2r\int_0^{\infty}K(\lambda,\,k,\,h)J_0(\lambda r)d\lambda\right] \qquad (4.7)$$

where $K(\lambda) = k\exp(-2\lambda h)/[1 - k\exp(-2\lambda h)]$ and J_0 is the Bessel function of order zero as shown by Stefanescu and Schlumberger (68).

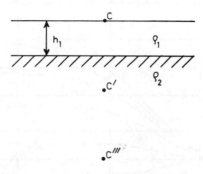

Fig. 29

With the knowledge of V it is possible to calculate ρ_a from equation (4.2) for any electrode configuration.

We shall confine ourselves to the Schlumberger array, equation (4.4). The ratio ρ_a/ρ_1 in this case is shown in Fig. 30 on a double logarithmic plot as a function of L/h_1 for values of ρ_2/ρ_1 from 0 (perfectly conducting substratum) to ∞ (perfectly insulating substratum).

It will be seen that ρ_a approaches ρ_1 when the current electrode separation is small compared to the thickness of the top layer and ρ_2 when it is large. The transition from ρ_1 to ρ_2 is, however, smooth and no simple general rule based on specific properties of the curve (e.g. the gradient) can be devised to find the thickness h_1.

With the addition of a third layer (h_2, ρ_2) sandwiched between the top layer (h_1, ρ_1) and the substratum (ρ_3), the problem becomes much more complicated. The apparent resistivity curve can then take four basic shapes known as Q (or DH, descending Hummel), A (ascending), K (or DA, displaced anisotropic) and H (Hummel type with minimum), depending upon the relative magnitudes of ρ_1, ρ_2, ρ_3 (Fig. 31). In every case, however, ρ_a approaches ρ_1 for small values of L and ρ_3 for large ones. At intermediate values of L it is influenced by the resistivity of the middle layer.

Master curves such as those in Fig. 30 cannot be conveniently presented for the three-layer case in a single diagram. The examples in Fig. 31 will

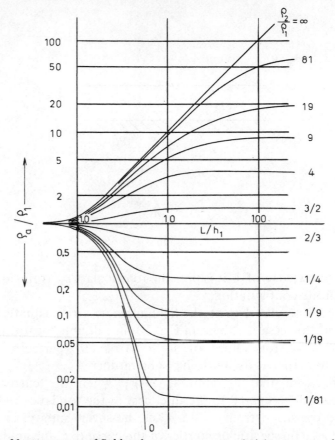

Fig. 30.   Master curves of Schlumberger apparent resistivity on a two-layer earth

nevertheless give an idea of the type of curves to be expected. Extensive catalogues have been published elsewhere (*69, 70*) and can be used with advantage in interpreting the field data.

The manner of using them or Fig. 30, provided that the data can be plausibly referred to the number of layers for which the master curves are constructed, is very simple. The observed ρ_a is plotted against $L = C_1 C_2 / 2$ on a transparent double logarithmic paper with the same modulus as the master curves paper. Keeping the respective axes parallel, the transparent paper is slid on various master curves in succession until a satisfactory match

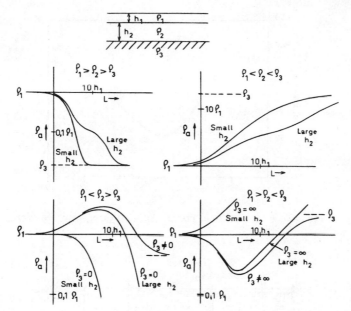

Fig. 31.   Q(or DH), A, K (or DA) and H type curves in vertical electrical sounding (VES)

is obtained with some curve (if necessary an interpolated one). The value of $C_1 C_2$ coinciding with the point 1.0 on the x-axis of the matching master curve gives h_1 and the value of ρ_a coinciding with the point 1.0 on the y-axis gives ρ_1. The values h_2, ρ_2 ... etc. are obtained from the appropriate parameters belonging to the matching master curve. The h and ρ values thus obtained are, however, subject to the principle of equivalence (p. 90).

The validity of the above procedure is easily grasped when one notes that either side of (4.6) or (4.7), for example, can be expressed in a dimensionless form.

It is noted that there is again no simple relation between the coordinates of the turning points or the inflection points of the ρ_a curves and the parameters h_1, h_2 and ρ_2.

A typical three-layer field curve is shown in Fig. 32a. It was obtained on a large lake with a fairly uniform depth of about 5 m. An interpretation made by comparison with the three-layer master curves gives $h_1 = 4$m, $\rho_1 = 340$ ohmm, $h_2 - 22$ m, $\rho_2 = 38$ ohmm and $\rho_3 = \infty$. Samples of the lake water

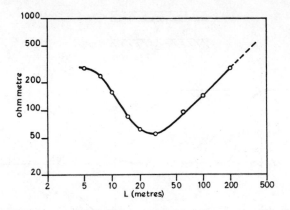

Fig. 32a.   A VES field curve (Boliden Company, Sweden)

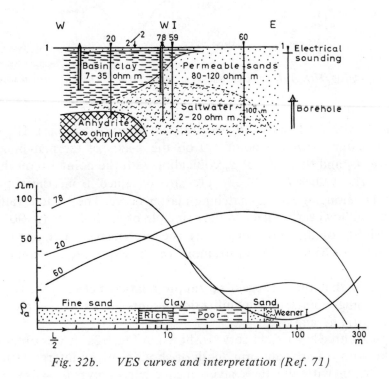

Fig. 32b.   VES curves and interpretation (Ref. 71)

(*L/2* in this figure has the same significance as *L* in section 4.3)

86

yielded a mean resistivity of 295 ohmm in the laboratory. Subsequent drilling has shown the second layer to consist of 21.3 m thick clayey bottom sediments on a granite basement. It is underlain by a badly conducting crystalline bedrock.

Another example of electrical sounding is shown in Fig. 32b together with the geological section (*71*). The object of this investigation was to locate permeable sands bearing fresh water. The interpretation of the sounding graphs, which to a large extent led to the geological picture shown, is not easy although certain qualitative correlations will be readily recognized.

4.5 The kernel function of resistivity

The potential distribution on the surface of an earth composed of four or more layers is often needed. It can be determined by electric images but the numerical computations using conventional methods are laborious. A rigorous and accurate method for rapidly calculating the apparent resistivity curve for any combination of thicknesses and resistivities in a stratified earth having a perfectly insulating or perfectly conducting substratum, and having strata thicknesses rational to each other, was developed by Flathe (*72*) but on account of the inherent limitations, the growing number of additions to master curves, access to computers, and other developments about to be described, the method is now largely of historical importance.

If in (4.5) σ (=$1/\rho$) is constant we get Laplace's equation $\nabla^2 V = 0$ which is true within each layer of the stratified earth. The potential distribution on the surface is again given by the same equation as (4.7) but with $K(\lambda)$ a function of all the strata thicknesses and resistivities. $K(\lambda)$ is known as the kernel function of resistivity. Its form for a two-layer earth has already been mentioned. For a three-layer earth, $K(\lambda) = [k_1 \exp(-2\lambda h_1) + k_2 \exp\{-2\lambda (h_1 + h_2)\}]/[1 - k_1 \exp(-2\lambda h_1) - k_2 \exp\{-2\lambda (h_1 + h_2)\} + k_1 k_2 \exp(-2\lambda h_2)]$. For an arbitrary, horizontally stratified earth $K(\lambda)$ is obtained by solving a certain system of linear simultaneous equations resulting from the boundary conditions of the problem (*74b*).

From (4.4) and (4.7) we get, for the Schlumberger array,

$$\rho_a = \rho_1 \left(1 + 2r^2 \int_0^\infty K(\lambda) J_1(\lambda r) \lambda \, d\lambda\right) \qquad (4.7a)$$

since $J_0'(x) = -J_1(x)$ where J_1 is the Bessel function of order 1.

Now, according to Hankel's transformation in Bessel function theory, if we have a function f(r) such that,

$$f(r) = \int_0^\infty K(\lambda) J_n(\lambda r) \lambda d\lambda$$

then

$$K(\lambda) = \int_0^\infty f(r) J_n(\lambda r) r dr$$

Applying the transformation to (4.7a) we find that

$$K(\lambda) = \int_0^\infty (1/2)\rho_1^{-1}(\rho_a - \rho_1) r^{-1} J_1(\lambda r) dr \qquad (4.7b)$$

This equation, first observed by King (*63*), shows that the kernel function can be computed unambiguously from a measured ρ_a curve (in the Schlumberger sense) by numerical integration and implicitly contains all the information about the layered earth. It plays a central role in the modern theory of the interpretation of resistivity data (p. 89).

4.6 Direct derivation of the layered earth parameters

The electric conductivity of a layered earth is a function of the depth only. According to the Slichter-Langer theorem, therefore, the knowledge of the surface potential due to a point electrode should suffice to determine the thicknesses and resistivities of the various layers. Two interpretation methods based on this approach are available.

The first method can be traced back to Hummel (*67*) although it has been subsequently refined and improved by several workers (*73, 74, 85*). Let us take the three-layer case (ρ_1, ρ_2, ρ_3) of Fig. 31 with $h_2 > h_1$. Clearly, as long as the current electrode separation does not exceed a certain value the apparent resistivity curve will not differ appreciably from a two-layer case with the same ρ_2/ρ_1 as that in the three-layer case under consideration. At larger electrode separations the third layer, that is, the infinite substratum (ρ_3) will influence the measurements. Hummel showed that for sufficiently large separations the apparent resistivity curve obtained is virtually the same as that for a two-layer case with the same substratum but with a top layer of thickness $\bar{h} = h_1 + h_2$ and a resistivity given by $\bar{h}/\bar{\rho} = h_1/\rho + h_2/\rho_2$. It will be noticed that $\bar{\rho}$ is derived by applying Kirchhoff's law for resistances in parallel.

Thus by matching the initial branch of a measured resistivity curve with an appropriate curve in Fig. 30 we obtain ρ_2 and h_1 (ρ_1 is known from the asymptote of ρ_a for very small electrode separations). Similarly, by matching the branch obtained with large electrode separations we get $\rho_3/\bar{\rho}$ and $\bar{h}$ again with the help of Fig. 30. Then ρ_3 and h_2 can be separately evaluated.

The above procedure is facilitated by the use of an auxiliary diagram in which $1 + (h_2/h_1)$ is plotted along the x-axis and $\bar{\rho}/\rho_1$ on the y-axis, both on a logarithmic scale, and for this reason is known as the auxiliary curve (or auxiliary point) method.

The method is easily extended to any number of layers with thicknesses $h_1, h_2, h_3, \ldots$ and resistivities $\rho_1, \rho_2, \rho_3, \ldots$. It is, of course, valid only if $h_1 < h_2 < h_3 \ldots$ but can also be validly employed in many other cases by constructing special auxiliary diagrams (*73, 85*).

The second direct method of interpretation is due to Pekeris (*76*) and is based on Slichter's analysis. We shall merely indicate the practical procedure; the theoretical justification will be found in Pekeris' original paper.

It is easy to measure the surface potential $V(r)$ of a single point electrode by removing the other current electrode to great distance. We compute the function

$$k(\lambda) = \lambda \int_0^\infty V(r)J_0(\lambda r)r\,dr \qquad (4.8)$$

by numerical or mechanical integration and plot in $|f_1(\lambda)|$ against λ where

$$f_1(\lambda) = \frac{k(\lambda) + 1}{k(\lambda) - 1} \qquad (4.9)$$

For large λ the points will lie on a straight line with a slope $2h_1$ and an intercept $\ln(1/k_1)$ with the y-axis where $k_1 = (\rho_2 - \rho_1)/(\rho_2 + \rho_1)$. If all the points lie on a straight line $h_2 = \infty$ (two-layer case); otherwise with h_1 and k_1 thus determined we proceed to calculate another closely similar function $f_2(\lambda)$, the plot of whose logarithm gives in turn h_2 and $k_2 = (\rho_3 - \rho_2)/(\rho_3 + \rho_2)$.

The process is continued by calculating a third function f_3 if all the points do not lie on a straight line, that is, if $h_3 \neq \infty$ and so on.

This method implies no restrictions on the relative magnitudes of $h_1, h_2, h_3, \ldots$ and is thus quite general. Unfortunately it requires considerable computation.

However, Koefoed has recently developed the method into a rapid practical procedure (74b). He starts by constructing $K(\lambda)$ from the observed ρ_a curve (4.7b) by means of a small number of standard curves. Next he defines a modified kernel function $G_n(\lambda) = K(\lambda)/\{1 + K(\lambda)\}$ and this he treats in essentially the same manner as $f(\lambda)$ above to obtain successive strata parameters.

Principles of equivalence and suppression.
The brief description of the above two methods has been given merely to indicate the theoretical scope of earth resistivity measurements. In the actual application of these methods to a particular field problem, the most serious limitations are those set by the maximum distance from the current source to which the electric field is given, and by the irregularities in the field due to surface non-homogeneities.

On account of these two causes and the finite accuracy of all measurements, widely different resistivity distributions may lead to ρ_a curves which, although they are not identical, cannot be distinguished from each other in practice. This introduces an ambiguity in the interpretation. Consider, for example, a relatively thin layer sandwiched between two layers whose resistivities are much higher than that of the sandwiched layer. The current flow in the earth will then tend to concentrate into the middle layer and evidently the fraction of the total current carried by it will be unaltered if we increase its resistivity (ρ) but at the same time increase the thickness (h) in the same proportion. Thus all the middle layers for which the *ratio h/ρ* is the same (within certain limits on h and ρ) are electrically equivalent in this case. On the other hand, if the resistivity of the middle layer is much larger than that of the layers on either side, the electric current density in it will tend to be less than in the layer below. The fraction of the total current that penetrates through the middle layer to the deeper level is impelled by the electromotive force across the layer, which will evidently be unaltered if we increase the thickness of the middle layer provided we, at the same time, decrease its resistivity in the same proportion. Thus, in this case, all middle layers for which the *product $h\rho$* is the same are electrically equivalent so that, again, h and ρ cannot be separately determined. Similarly, in the first case, if the thickness of a layer is very small compared to its depth (and its resistivity is non-zero) its effect on the ρ_a curve is so small that the presence of the layer will be suppressed.

Vertical discontinuities.

The case of two homogeneous rock formations separated by a plane vertical boundary (Fig. 33a) has been treated by several writers (*77, 78, 79*). If we have a single point electrode C, the other electrode being at infinity, we get for the apparent resistivity defined by equation (4.4).

$$\rho_a = \rho_1\left[1 - k_1\,\frac{r^2}{(2a-r)^2}\right] \quad r < a \text{ (medium 1)}$$

$$\rho_a = \rho_1(1 + k_1) \qquad\qquad r > a \text{ (medium 2)}$$
(4.10)

with k_1 as in the last section. Typical curves are given in Fig. 33a.

It will be seen that the apparent resistivity is discontinuous at the boundary. The discontinuity will be evident in practice as a steep gradient of the resistivity curve (Fig. 33b).

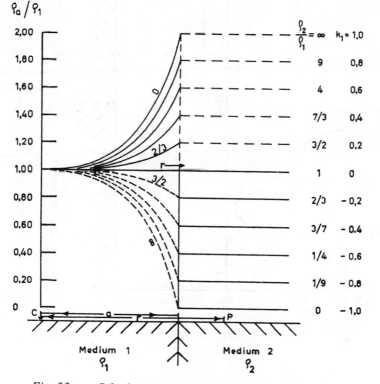

Fig. 33a.   Calculated Schlumberger ρ_a curves (Ref. 79)

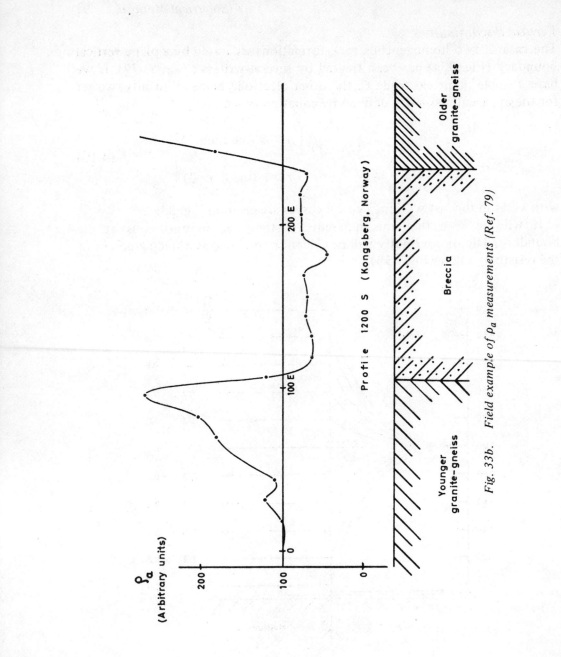

Fig. 33b. *Field example of ρ*_{*a*} *measurements (Ref. 79)*

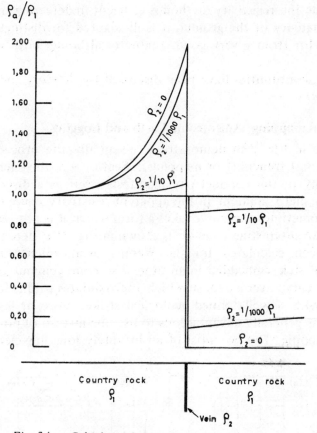

ρ_a/ρ_1

Fig. 34.   *Calculated Schlumberger ρ_a across a vein*

In Fig. 34 are shown the apparent resistivity curves in another commonly encountered case, namely a thin conducting dike or vein cutting through the surrounding rock.

Figures 33a and 34 incorporate two features which are worth a special mention since they apply to all work with the resistivity methods. Firstly, small resistivity contrasts cause comparatively large departures of the ratio ρ_a/ρ_1 from 1.0. Secondly, the ratio is practically unchanged whether the resistivity contrast is moderately large, say $\rho_2/\rho_1 = 10$ or very large, say $\rho_2/\rho_1 = 10,000$. This is often known as the 'saturation effect'. Consequently, while the resistivity method is efficient in detecting small variations in the conductivity of the ground, it is ill-adapted for distinguishing, say, a good conductor from a very good conductor although both will be readily detected.

Dipping discontinuities have been discussed by de Gery and Kunetz (*80*), and Maeda (*81*).

4.7 Electrical mapping, Anisotropic Earth and Logging

The example in Fig. 33b demonstrates essentially the procedure that was termed electrical trenching or mapping in Section 4.3. In general, any lateral inhomogeneity in the conductivity manifests itself as a discontinuity or a more or less sharp gradient in the apparent resistivity curve or its distance derivative. Sometimes the curve takes a form which it is not easy to visualize intuitively. An interesting example is shown in Fig. 35a. Here the theoretical curve has been calculated for the Wenner arrangement taken across a hemispherical sink embedded in an otherwise homogeneous ground. Figure 35b is a field curve over a deposit which approximates to this form (*82*).

In areas with a well-defined geological strike, more or less constant in direction, it is generally advantageous to use line instead of point sources for electrical mapping. The potential of an infinitely long line electrode is given

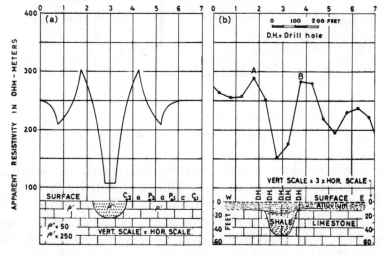

Fig. 35.   Wenner ρ_a across hemispherical sinks (Ref. 82)

by $(I\rho/\pi)\ln r$ where I is the current per unit length. For a finite electrode the logarithmic term is somewhat more complicated (*94*).

Figure 36 shows the results of an electrical mapping survey in an ore-bearing area using parallel electrodes, 200 m long and 1,200 m apart. The potential probes (40 m apart) were moved along lines perpendicular to the current electrodes and between them. The country rock in the region, with

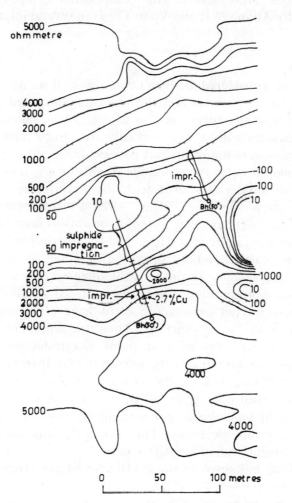

Fig. 36.   Resistivity map of an area in North Sweden (Boliden Company, Sweden)

apparent resistivities less than 1,000 ohmm, is more or less uniformly impregnated with pyrite and pyrrhotite with mineable concentrations of chalcopyrite in places. The boundaries between the barren and the impregnated rock are clearly indicated by the steep gradient in the resistivity curve.

The electric conductivity distributions considered in this chapter are rather simple cases. More general and complicated conditions have been treated recently by Alfano (*83*) and Vozoff (*84*) in several interesting papers.

Anisotropic earth.
Although we have considered homogeneous as well as non-homogeneous earth it has been tacitly assumed that all the media concerned are electrically isotropic. However, many rocks such as shales, slates, laminated ores and even some bedrock formations are markedly anisotropic and serious errors can arise in the interpretation if this fact is neglected.

We must distinguish between two kinds of anisotropy: a microanisotropy in which the individual grains of a rock are anisotropic owing to their structure and a mega-anisotropy arising when a given formation contains several different facies (themselves isotropic or anisotropic) alternating in a regular fashion. In general the two effects will be superimposed.

The anisotropic earth has been discussed in detail by Schlumberger et al. (*87*) and by Maillet (*86*). The main consequences of anisotropy bearing on the earlier discussions in this chapter may be stated as follows.

Let a homogeneous but anisotropic earth have resistivities ρ_l and ρ_t, parallel and transverse to its surface. Then the apparent resistivity ρ_a determined by a linear array of four point electrodes on the surface is $\sqrt{(\rho_l \rho_t)}$ while the apparent resistivity perpendicular thereto measured in a vertical borehole will be paradoxically enough, $\rho_l (< \sqrt{(\rho_t \rho_l)})$.

Secondly, any anisotropic layer of thickness h can be replaced by an isotropic layer of thickness $h\sqrt{(\rho_t/\rho_l)}$ without altering the surface potential distribution of a point electrode. This evidently introduces an indeterminacy in the interpretation of resistivity data.

The ratio $\sqrt{\rho_t/\rho_l}$ is known as the coefficient of anisotropy. For layered rocks such as shales, gneisses etc. it is usually of the order 1.10–1.30.

Electrical logging.

Apparent resistivity determinations can be made by sinking an electrode array in a borehole. In practice one current electrode is often placed on the surface at a large distance from the borehole while the other current electrode and the two potential probes, with fixed mutual distances, are lowered in the hole, the electrodes being pressed against the walls by springs. A similar technique is also used for measuring self-potentials in boreholes.

These procedures of electric measurements in boreholes are known as electrical logging or coring. They have been used chiefly in oil fields for stratigraphic correlations from one borehole to another, it being found that one and the same geological formation within a restricted area is almost always associated with characteristic resistivity values or self-potentials (*62, 87*).

The SP values in sedimentary formations are almost entirely concentration potentials or potentials arising due to diffusion of ions. It has been claimed (*75*) that weak formation-potentials can be enhanced by the addition of certain chemicals to the drilling mud. The subject of electric logging has been treated in great detail by Lynch (*170*).

4.8 The resistivity of rocks and minerals

The magnitude of the electric anomalies on a non-homogeneous earth depends upon the resistivity differences between different rocks, or, more correctly, upon factors of the type $(\rho_2 - \rho_1)/(\rho_2 + \rho_1)$.

The resistivity of rocks is an extremely variable property ranging from about 10^{-6} ohmm for minerals such as graphite to more than 10^{12} ohmm for dry quartzitic rocks. Most rocks and minerals are insulators in the dry state. In nature they almost always hold some interstitial water with dissolved salts and therefore acquire an ionic conductivity which then depends upon the moisture content, the nature of the electrolytes and their concentration. The form of the pores in a rock plays a subordinate role in determining the conductivity.

Some minerals, notably graphite, pyrrhotite, pyrite, chalcopyrite, galena and magnetite (a ferrite!) are relatively good (electronic) conductors (Table 6). A dissemination of such minerals within a rock can also make it a

TABLE 6

Electric resistivities (ohmm)

Rocks and sediments		Ores	
Limestone (marble)	$>10^{12}$	Pyrrhotite	$10^{-5}-10^{-3}$
Quartz	$>10^{10}$	Chalcopyrite	$10^{-4}-10^{-1}$
Rock salt	$10^{6}-10^{7}$	Graphite shales	$10^{-3}-10^{1}$
Granite	$5,000-10^{6}$	Pyrite	$10^{-4}-10^{1}$
Sandstones	$35-4,000$	Magnetite	$10^{-2}-10^{1}$
Moraine	$8-4,000$	Haematite (†)	$10^{-1}-10^{2}$
Limestones	$120-400$	Galena	$10^{-2}-300$
Clays	$1-120$	Zinc blende	$>10^{4}$

† Stoichiometric Fe_2O_2 Insulator

better conductor. Others, such as zinc blende, are also electronic conductors but very poor such at ordinary temperatures.

Practically all rocks and minerals are semiconductors, their resistivity decreasing with increasing temperature. Especially in the sulphide minerals, donor and acceptor impurities play a great part in determining the absolute magnitude of the electric conductivity. The activation energies are however poorly known for most of the naturally occurring sulphides and oxides.

The resistivity of porous, water-bearing rocks (free of clay minerals) follows Archie's law, $\rho = \rho_0 f^{-m} s^{-n}$ where ρ_0 is the resistivity of the water filling the pores, f is the porosity (volume fraction pores), s is the fraction of pore space filled by the water and n, m are certain parameters. The value of n is usually close to 2.0 if more than about 30 percent pore space is water-filled but can be much greater for lesser water contents.

The value of m depends upon the degree of cementation or, as this is often well correlated with geologic age, upon the geologic age of the rock. It varies from about 1.3 for loose, Tertiary sediments to about 1.95 for well-cemented Palaezoic ones, but can be outside this range for individual formations (*110*).

4.9 Induced polarization

If an electric current in the ground is intercepted the voltage across P_1, P_2 (Fig. 28) does not drop to zero instantaneously. It is found instead to relax for several seconds (or minutes) starting from an 'initial' value which is a small fraction of the voltage (*V*) that existed when the current was flowing (Fig. 37a). This phenomenon has been termed *induced polarization* or *overvoltage*. That it occurs in dielectrics in the laboratory has bccn known for a long time (*88*).

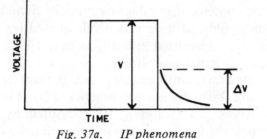

Fig. 37a.   IP phenomena

When measurements are made by passing d.c. pulses of, say, 20 s duration the magnitude of IP is often expressed as $\Delta V/V$ (millivolt per volt or per cent) where ΔV is the overvoltage remaining at a definite time t, say 5 s, after cut-off. Sometimes the area under the decay curve is also used to express IP as millivoltsec per volt. The ratio $\Delta V/V$ is independent of V (*89*).

It takes a finite, although short, time (probably <0.5 s) to produce an overvoltage. This means that, for an uninterrupted current flow the (complex) impedance of the ground must depend on the frequency of the current as is, in fact, borne out by observation. Hence, IP measurements can also be made with a.c. It is then possible to express the magnitude of the effect as $Z(0)/Z(f)$ where $Z(0)$ is the d.c. (or very low frequency) impedance and $Z(f)$ the impedance at frequency f. Frequencies of the order $0.01-10$ c/s are used.

The overvoltage effects resemble ordinary dielectric behaviour; however, there are important differences. The relaxation is not a simple exponential one but rather a hyperbolic one in many cases. Also the decay constant in IP, about a fraction of a second, is $3-4$ orders of magnitude greater than the *RC* constant of the ground. Moreover the dispersion referred to above occurs at audio and sub-audio frequencies where displacement currents are totally negligible.

The d.c. and a.c. methods give essentially the same information being related to each other through the Fourier representation of the d.c. pulse. As an approximate relation Marshall and Madden (*90*) state that the percentage decrease in impedance at frequency f corresponds to 0.1 mV/V-value at $t = 1/2\pi f$.

The parameter $2\pi \times 10^5 \times [Z(0) - Z(f)]/Z(0)Z(f)$ has often been used to characterize different media and termed the 'metal factor' because of its correlation with the content of conducting metallic minerals in the ground, but the correlation is only of a very general, qualitative nature. It ranges from less than 1 for unmineralized granites to over 10^3 mho/m for massive sulphides. The factor $2\pi \times 10^5$ is arbitrary.

Physically, the metal factor is nothing but the difference between the admittances $Y(f)$ and $Y(0)$. In normal practice $Y(f)$, $Y(0)$ are simply the 'apparent conductivities', $\sigma_a(f)$ and $\sigma_a(0)$, measured by any one of the innumerable electrode configurations possible (section 4.3). Their difference depends on the configuration used. In recent years attention has also been turned to measuring the phase lag of the voltage ΔV, with respect to the current (*171, 172*) besides its magnitude. Maximum lags, in practice, are usually a few hundredths to a tenth of a radian at a frequency of, say, 1 c/s. These imply delay times of the order of 1−15 ms.

Origin of overvoltages.
The electric conduction paths in the ground are normally ionic but they may be sometimes hindered to a greater or less extent by mineral particles (e.g. pyrite grains) in which the carriers of current are electrons.

It is well known that when a current passes across a metal electrode (electronic conductor) dipped in an electrolyte, charge can pile up at the interface when all the processes in the electrolytic reaction are not equally rapid. This produces the familiar back e.m.f. or electrode polarization. There can be little doubt that the IP observed over sulphide ores is a manifestation of such 'electrode' polarization. The effect will be enhanced if the ore is disseminated rather than compact since it is essentially a surface phenomenon and the polarization will be large owing to the large total surface of the particles. Values of more than 10 per cent for $\Delta V/V$ are not uncommon on such ores (Fig. 64).

Overvoltages are, however, observed even in the absence of conducting minerals. The presence of clay particles appears to be a necessary condition for this so-called 'normal' effect for it is not observed on clean quartz sand.

It must be realized that the electromagnetic coupling between the current cables and the ground will also give a signal similar to the IP signal. This signal can, however, be avoided if the cable layout is properly selected and if very low frequencies are used.

The possible causes of the normal effect have been exhaustively discussed by Marshall and Madden (90), Vacquier and others (91) and Mayper (92). It is evident from their analyses that factors such as electro-osmosis, thermoelectric couplings and surface electronic levels on insulating minerals are unlikely to be important. The overvoltage effect is most probably due to ionic exchanges and the setting-up of diffusion potentials somewhat as follows.

An electric current alters the distribution of adsorbed ions on the clay particles present in the pores of a rock. Some lattice ions may also be removed by electrodialysis, that is, by the straightforward process of replacing the ions by hydrogen ions. The return of the clay particles to equilibrium will appear as an overvoltage response.

In general the electrode polarization and the 'normal' effect will be superimposed and complicate the interpretation.

The tendency of the ions to diffuse back to their equilibrium positions constitutes an impedance to the passage of the current. This is known as the Warburg impedance in electrochemistry. In the frequency domain the Warburg impedance causes a phase lag to appear between the injected current and the measured voltage.

Suppose the ions in the electrolyte obey the diffusion equation $\partial C/\partial t = D(\partial^2 C/\partial x^2)$ where C is the ion concentration and D the diffusion coefficient. It is well known that the solution of this equation is of the form $C = C_0 \exp(-x\sqrt{\pi v/D}) \sin(2\pi v t - x\sqrt{\pi v/D})$ which shows that the amplitude and the phase of the overvoltage response should depend upon $\sqrt{v}$ if v is the frequency of the current. However, from the results of accurate experiments (172) it appears that the dependence is of the type $v^{-\alpha}$ where $\alpha \neq 1/2$. Simple diffusion is thus not sufficient to explain the IP phenomenon. This is also evident from the fact that the time constants and phase lags derived from the ~~Warburg impedance theory~~ (93) are several orders of magnitude smaller than those observed in practice.

In some cases the IP method seems to afford a greater resolution of anomalies than the resistivity method and is at present being applied in ore prospecting on a very large scale. The greater resolution can sometimes be due to variations in the total area presented by the mineral or clay particles in the ground and not to variations in their content as such.

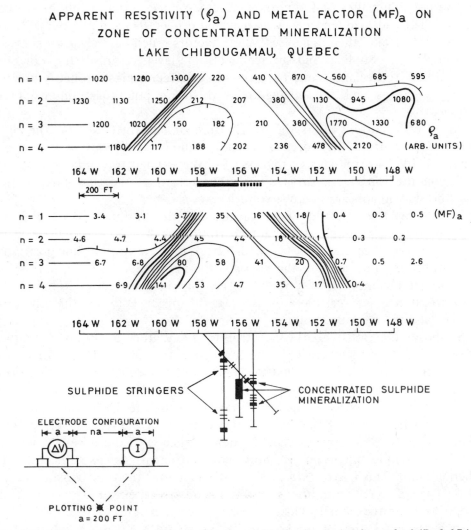

Fig. 37b.   *A frequency domain IP survey with the dipole-dipole method (Ref. 174)*

Fig. 37b is an example of a ρ_a and IP (frequency domain) survey (*174*). It should be evident from p. 99 that an IP measurement necessarily involves a concomitant resistivity measurement. The measurements in Fig. 37b were made by using a number of dipole-dipole arrays (Fig. 28) with different n values. The manner of plotting the readings should be clear from the inset. A diagram obtained in this manner is often qualitatively regarded as representing an 'electric vertical section' through the profile, since increasing n values give information from increasingly larger depths but the plot must not be taken too literally. Vertical tabular bodies of constant thickness, for example, can be shown to give wedge-shaped anomalies on this plot. Thus the patterns in Fig. 37b do not by any means necessarily indicate a broadening of the mineralization towards the deeper levels.

5 · *Electromagnetic continuous wave, Transient and Telluric Methods*

5.1 Introduction

If an electromagnetic field is produced on the surface of the ground currents will flow in subsurface conductors in accordance with the laws of electromagnetic induction. These currents give rise to secondary electro-magnetic fields which distort the primary field at any point on the surface. In general, the resultant field, which may be picked up by a suitable search coil, will differ from the primary field in intensity, phase and direction and reveal the presence of the conductors.

If the primary field is transient the secondary currents and their field will decay gradually when the primary field has ceased to exist. The decay is faster the higher the resitivity of the medium in which the currents flow. In this case we cannot talk of any unique phase relations since the signal contains an infinite number of frequencies. The discussion of transient methods will be deferred until section 5.9.

A great advantage of the electromagnetic methods is that they can be successfully applied even when conductive ground connections indispensable for the methods of the last chapter cannot be made owing to highly resistive (or insulating) surface formations. This is frequently the case in arid tracts or in the polar and sub-polar regions where the ground may be frozen to a considerable depth.

On the other hand one of the troublesome effects in the electromagnetic methods is that the secondary currents in superficial layers of good conductivity, e.g. clays, graphitic shales, etc. may screen the deeper conductors partially or wholly from the primary field. The latter which are the real objects of exploration will then produce weak or no distortions (anomalies) in the primary field and may therefore be indetectable.

5.2 Phase relations

It will be convenient at the outset to recall briefly a few elementary ideas.

It is well known that if a primary magnetic field $P = H_0 \sin \omega t$ of frequency $\omega/2\pi$ acts on an electric circuit, say a coil, the secondary induced e.m.f. lags $\pi/2$ behind the primary field. If Z and L denote the resistance and the self-inductance of the coil, the current in the coil and the secondary magnetic field produced by it lag $\pi/2 + \phi$ behind the primary field, where $\phi = \tan^{-1}(\omega L/Z)$, the lag $\pi/2$ being caused by the fundamental law of induction and ϕ by the properties of the secondary circuit.

The relation between the primary (P), the secondary (S) and the resultant (R) fields is shown in Fig. 38a which is a conventional vector diagram of electric circuit theory with the only difference that the axis of ordinates is positive downwards.

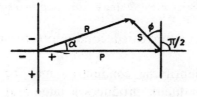

Fig. 38a. Phase relations

It follows that a very good conductor produces a secondary field almost opposite in phase to the primary field ($Z \to 0, \phi \to \pi/2$) while a bad conductor produces a field that lags 90° behind the primary one ($Z \to \infty, \phi \to 0$). The same phase effects are obtained on increasing and decreasing the frequency of the primary field (or the inductance in the secondary circuit).

The component of S in phase with P (the real component) is $- S \sin \phi$; the component lagging 90° behind P (out-of-phase or imaginary component) is $+ S \cos \phi$.

The real and imaginary components of the magnetic field can be conveniently expressed in units such as millivolts, nano-Wb/m² per ampere primary current, etc. but it is also customary to express them as fractions or percentages of the primary field.

The real and imaginary components of the secondary response of a single-turn loop, to a homogeneous sinusoidal field, are plotted in Fig. 38b against Z/ω. This case reproduces qualitatively every essential detail of the

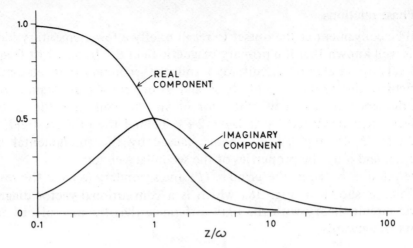

Fig. 38b.   *Response of a single-turn loop to a sinusoidal field*

induction response of complicated conductors to sinusoidal fields, homogeneous or otherwise.

Simple rules for identifying conductors may be formulated from Fig. 38b. Thus, a good conductor produces a large real but a small imaginary component while a bad conductor produces a relatively large imaginary but a small real component. If the conductor has a 'medium' resistivity both the components are moderately large. Quantitatively, the ratio of magnitudes 'Real/Imaginary' is often used this being greater than 1 for good and less than 1 for bad conductors.

Elliptic polarization.

Suppose that the plane of the coil in the above example is vertical and that the (homogeneous) primary field is horizontal. The resultant field at any point above the coil may be resolved into a horizontal (X) and a vertical (Y) component. It can be seen that

$$X = H_0 \sin \omega t + k \sin(\omega t - \pi/2 - \phi) = A \sin \omega t + B \cos \omega t$$

$$Y = k' \sin(\omega t - \pi/2 - \phi) = A' \sin \omega t + B' \cos \omega t \qquad (5.1)$$

where k, k' are the amplitudes of the horizontal and the vertical components of the secondary field.

Eliminating ωt from (5.1) we get

$$(A'^2 + B'^2)X^2 + (A^2 + B^2)Y^2 - 2(AA' + BB')XY$$
$$- (A'B - AB')^2 = 0 \qquad (5.2)$$

which is the general equation of an ellipse. This means that the resultant field at any point is elliptically polarized, the vector $\sqrt{(X^2 + Y^2)}$ describing an ellipse $\omega/2\pi$ times per second. In fact, the resultant field is always elliptically polarized irrespective of the nature of the primary field or the number or nature of the secondary circuits.

The ellipse will degenerate into a straight line if the conductor is very good.

Classification of electromagnetic methods.
Many early electromagnetic methods were based upon determining the azimuth and the dip of the ellipse of polarization as well as its major and minor axes which, it can be shown, determine the amplitudes of the real and imaginary components of the resultant field. It is easy, in principle, to find these parameters by means of a search coil which can be oriented in any desired direction.

If the coil is connected via an amplifier to a null detector the plane of the ellipse will be indicated by the position of the coil in which a null is obtained in the detector. Setting the coil normal to the plane of polarization and measuring the maximum and minimum signals induced in it as it is turned through one complete resolution, one obtains the two axes of the ellipse.

Although the method provides the most complete information about the electromagnetic field at a point it is cumbersome to use and the desired accuracy is not easy to attain.

In a number of other methods (the so-called tilt angle methods) only the tilt of the major axis of the polarization ellipse from the horizontal or the vertical is determined. However, tilt angles do not provide complete information about the electromagnetic field. Some of the recently devised methods (e.g. Afmag, VLF etc.), discussed later, are also essentially tilt-angle methods.

The electromagnetic methods most commonly employed at present were perfected between about 1925 and 1940, chiefly in Sweden, although important pioneering work was also done in Australia (*77*) and America (*56*). There are two main categories:

1. Methods in which the source of the primary field is stationary and the receiver arrangement mobile and

2. Methods in which the source as well as the receiver is mobile.

The first category includes the so-called Compensator and Turam methods; the second includes methods or to be more accurate, measuring outfits, which go under various names such as Slingram, electromagnetic gun (EMG), VEM, HEM, etc., which are all basically alike.

5.3 The Compensator or Sundberg method

The primary layout in this method consists of a straight cable, some 400 to 4,000 (or more) metres long, grounded at both ends, through which an alternating current of low frequency ($< 1,000$ c/s) is passed. A large horizontal loop, usually rectangular, may also be laid on the ground instead of the cable.

The cable (or the long side of the primary loop) is generally placed approximately parallel to the geological strike in the area and the electromagnetic field is investigated at regular intervals along lines perpendicular to it. In the case of the loop, observations can be made inside as well as outside the loop.

For many, if not most purposes, it would be sufficient to measure the amplitude and phase of either the vertical or the horizontal component of the resultant field but measurements of both may be called for in detailed work. A search coil consisting of several turns of copper wire on a suitable frame is held horizontally (to measure the vertical component) or vertically (to measure the horizontal component) and the voltage induced in it is compared with a reference voltage. The latter is obtained from an auxiliary ('feeding') coil stationed near the primary layout. A method has also been tried in which the reference voltage is instead transmitted by an UHF carrier-wave modulated at the frequency of the primary current.

The comparison of voltages is made on a compensator, essentially an a.c. potentiometer, in which the inclusion of a reactive element enables one to determine phase differences. A simple circuit of this type is shown in Fig. 39. The real component is balanced by the voltage across the resistor R and the imaginary one by that across the variometer I, the balance being indicated by a silence in the telephones.

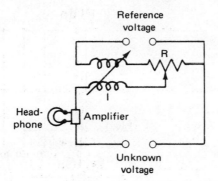

Fig. 39. A.c. compensator

The general problem of delineating subsurface conductors by electro-magnetic methods is one of locating secondary current concentrations. Owing to the skin-effect the currents tend to be localized on the boundaries of a conductor.

The most pertinent case, especially in such applications as ore prospecting, is that of a vertical sheet-like conductor representing an ore vein. Depending upon the position of such conductors with respect to the primary source they may be mapped according to the following considerations.

(1) If the vertical conducting sheet is cut transversely by the field of a long cable or that of a large loop the currents will evidently flow mainly along vertical surfaces, being strongest on the side nearest the primary source. The main features of the secondary magnetic field on the surface may be visualized by replacing the secondary current distribution by a vertical sheet having only transverse magnetization.

(2) If the sheet is *inside* a loop it will be subjected to a more or less uniform vertical field and the secondary currents will flow mainly in horizontal planes, being strongest on the upper face and decreasing downwards in depth. The secondary, surface magnetic field will resemble the field of a vertically magnetized sheet in its broad features.

Thus, the conductor will be located below the inflexion point of the secondary vertical field in case (1) and below the maximum in case (2). Figure 40 shows the results of some laboratory model experiments (after Sundberg) corresponding to the first case. It will be seen that the phase also changes in a characteristic manner along the line of measurements and, moreover, that the width of the conductor is almost exactly equal to the

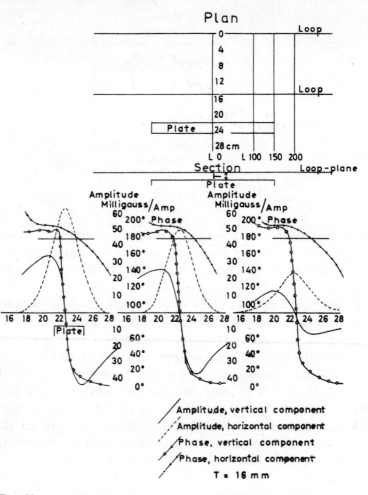

Fig. 40.   *Amplitude and phase anomalies in the compensator method*

distance between the points at which the phase of the vertical and horizontal components is 180°.

Perhaps the most interesting application of the compensator method is in the determination of the depth and conductivity of a number of horizontal conductors one below the other but not necessarily contiguous. A sedimentary column with interspersed conducting beds, for instance, can be approximated by such a model.

The electromagnetic field at a point P above a thin horizontal conducting sheet when an alternating current flows through an infinitely long cable, parallel to the sheet and at a height h above it, can be calculated from a formula derived by Levi-Civita (95). The field is a function of y, $h + z$ (Fig. 41) and the 'induction factor' of the conductor $p = (\frac{1}{2}) \, \mu_0 \times \omega d/\rho = 4\pi^2 \times 10^{-7} \, \nu d/\rho$ where ρ is the resistivity (ohm-m, d the thickness (m) of the conductor and $\nu = \omega/2\pi$ the frequency of the primary current. Evidently $1/p$ has the dimensions of a length if we recall that μ_0 has the dimensions henry/m = ohm-s/m.

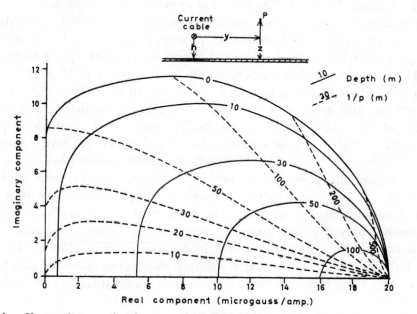

Fig. 41.   *Vector diagram for the secondary field due to a long current above a thin sheet conductor*

The real and imaginary parts of the vertical electromagnetic field above a horizontal conductor at a point distance 100 m from the primary current and at the same height as the current $(z = h)$ are plotted on the vector diagram in Fig. 41. Here the solid curves show how the components vary for different conductors with different $1/p$-values but at the same depth and the dotted ones how they vary for a conductor with a given $1/p$-value but situated at different depths. A line drawn from the origin to a point on a solid curve represents the total vertical field vector (cf. R in Fig. 38).

It will be realized that a single in-phase and out-of-phase observation at a known distance suffices to determine the depth as well as the induction factor of the thin horizontal conductor.

Another important property of the vector diagram in Fig. 41 is that it can be used to construct the electromagnetic field above any number of thin, plane and parallel conductors one below the other. The procedure may be illustrated by reference to the simple case of two conducting sheets 1 and 2, having induction factors p_1, p_2 placed at depths z_1 and z_2 ($> z_1$).

If the sheet 1 is immediately on top of sheet 2 the two form a 'combined' thin sheet at depth z_2 with an induction factor $p = p_1 + p_2$. The field of this sheet as well as the field of sheet 1 alone placed at depth z_2 can be read off Fig. 41. The vector difference between them gives the field of sheet 2 placed at depth z_2. This field can be added vectorially to the field of sheet 1 alone at depth z_1, also read off Fig. 41, and we obtain the field vector of the system of two thin sheets at depth z_1 and z_2. The theory of this construction and its generalization to any number of thin horizontal conductors have been discussed by Sundberg and Hedström (*96*).

In applying the above to deduce the unknown depths and induction factors of horizontal conducting beds from compensator observations the vector constructions are reversed. The total field vector is divided into its constituent parts and a solution obtained for the system of conductors. It should be realized that this is not a 'trial-and-error' procedure in the usual sense but one that gives a unique solution depending upon the number of observations available. This point has been very ably discussed by Sundberg and Hedström:

'Since every field vector measured means the determination of two coordinates in a vector diagram . . ., it will allow the determination of two unknowns, one depth and one induction factor. Every reading added, whether taken from another component of the field vector, at another distance from the primary cable or at another frequency, will therefore make it theoretically possible to determine two more unknowns. . . . In practical work it is necessary, however, in order to obtain a solution within reasonable time, to substitute a simple (but representative) system of conducting sheets, comprising not more than four or five unknown quantities. . . .'

The compensator method has been extensively used mainly for structural studies such as following up electrical 'key-beds' in oil field formations or following up the depth extension of flat-lying or gently dipping ore lenses (*61, 97, 98, 126*).

5.4 The Turam method

The compensator method requires a direct connection between the primary layout and the observation point and becomes therefore cumbersome when large areas are to be covered. This operational disadvantage is overcome in the Turam method devised by Hedström (*100*).

The primary field is produced as before by a long cable or a large loop, and two search coils, 10–50 m apart, are carried along the line of measurements. For each position of the coils, the ratio of the amplitudes of and the phase difference between the voltages induced in them are measured on a bridge type compensator, the former with an accuracy of about 0.01, the latter about 0.2°.

The coils are usually held horizontally (so as to compare the vertical components of the resultant field) but may sometimes be kept vertical with their planes either parallel or perpendicular to the measuring profile, while in some investigations one coil may be horizontal and the other vertical. We shall restrict the discussion to the case of two horizontal coils.

The quantities measured in Turam work are $V_1/V_2, V_2/V_3, \ldots$, etc. and $\alpha_2 - \alpha_1, \alpha_3 - \alpha_2, \ldots$, etc. where the V's are the amplitudes, and α's the phases of the vertical electromagnetic field at the stations 1, 2, 3. . . . To correct for the variation of the primary field (p) with the distance from the source the measured ratios are divided by the normal amplitude ratios p_1/p_2, $p_2/p_3, \ldots$, etc. This is particularly easy in the case of the long cable since the normal ratio of the vertical fields at any two points is simply the inverse ratio of the distances of the points from the cable, unless one or both of the coils are on a level different from the cable when an easily calculable correction must be applied.

The normalized or reduced ratios $V_1 p_2/V_2 p_1, V_2 p_3/V_3 p_2, \ldots$, etc. will all be equal to unity in the absence of subsurface conductors.

The normal phase differences are, of course, zero provided the ground is non-conducting. This statement is not strictly true because the phase of an electromagnetic field varies by 2π over a distance of one wavelength. However, the air-wavelength of the low frequency primary field used in geoelectric work is of the order of several hundred kilometres. This is very long compared with the distance, from the cable, of a kilometre or so up to which the measurements can be usually made so that the normal phase changes, although not strictly zero, are quite negligible. On conductive ground, however, the phase may change appreciably within relatively short distances. Moreover, the field will be elliptically polarized, the inclination of the ellipse becoming more and more horizontal away from the cable. In accurate and detailed work these effects must be taken into account as otherwise uncertainties will be introduced in the assessment of the normal field (*181, 182*).

It is evident that deviations of the reduced ratio from unity and the phase difference from zero indicate anomalous subsurface conditions.

Now, the phase difference measures essentially the horizontal gradient of the phase. As regards the reduced ratio it can be shown that, to a first approximation, it represents the function $1 - cs'/(1 + s)$ where s is the secondary field expressed as a fraction of the primary field at the point, s' its horizontal gradient and c the constant separation between the search coils. Thus, the *departures* of the reduced ratio from 1 constitute a measure of the horizontal gradient of the amplitude of the secondary field.

Typical vertical field indications obtained along three parallel Turam profiles across a sheet-like conductor are shown in Fig. 42. The conductor in question is a large pyrite ore body in an environment of porphyry rocks. It dips about 68° NE and the depth to its top surface is 14 m (*101*). The frequency of the primary field was 540 c/s in this work and the source was at 400NE. The marked asymmetry in the curves is probably due to the dip of the conductor. It is worth remembering that in case a sheet-like conductor has an attitude parallel to the primary field-lines at it, there will be no ratio and phase-difference anomalies above the conductor. Moreover, the signs of either indication will be reversed if the dip towards the source is shallower than this value, presuming throughout that the surrounding rocks are non-conducting.

It will be noticed that the ratio and phase difference curves show extremum values directly above the conductor and that they resemble the x-derivatives of the amplitude and phase curves for the vertical field in Fig. 40.

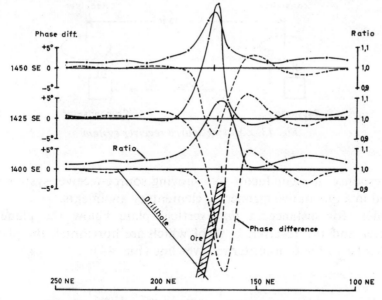

Fig. 42. Turam ratio and phase difference anomalies

5.5 The moving source and receiver method

The layout of this method is shown in Fig. 43. A battery-operated portable oscillator delivers an alternating voltage to a transmitter coil. The receiver is spaced at a fixed distance from the transmitter, usually between 25 and 100 m, and is connected to a compensator which is also fed by the reference voltage from the transmitter. The field acting on the receiver is measured in percent of the primary field present at it when the system is on electrically neutral ground.

Appropriate corrections to the compensator readings must be applied to take into account the changes caused in the primary field at the receiver, due to a change (e.g. on account of topography) in the mutual separation, or orientation of the coils. These corrections, which are only needed for the real component, are easily calculated since the primary field being an ordinary dipole field is known everywhere.

In principle, any mutual orientation of the transmitter and the receiver may be used but most surveys are carried out with both the coils either horizontal or vertical.

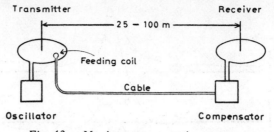

Fig. 43.   *Moving source-receiver system*

The response of a conductor to a moving source-receiver system is easily visualized in a qualitative manner by elementary arguments.

Consider, for instance, a *thin* vertical plate below the plane of the transmitter and the receiver, both of which are horizontal, the plate being perpendicular to the transmitter-receiver line (Fig. 44).

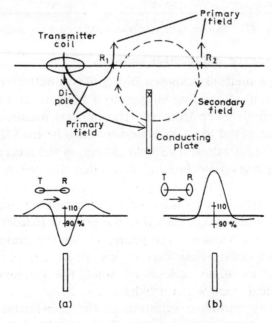

X – coordinate is the midpoint of the T–R line

Fig. 44.   *Origin of electromagnetic anomalies in the moving source-receiver method*

Suppose that at a particular instant the oscillating transmitting dipole is directed downwards so that the primary field cuts the plate as shown. The secondary currents in the plate will be into the plane of the figure at the top edge of the plate and out of it at the bottom. At any point R_1 on the same side of the plate as the transmitter the secondary field will be directed upwards, that is, in the same sense as the primary field at it. At a point such as R_2, however, the two fields will be in opposite directions and the resultant field will be less than the primary. Furthermore it is easy to see that if either the transmitter or the receiver is directly above the plate the field indicated by the compensator will be equal to the primary field.

Thus as the horizontal transmitter-receiver system with a fixed spacing is carried across the plate the field picked up by the receiver will have the appearance as in Fig. 44a. The maximum damping in the field (negative anomaly) is obtained when the midpoint of the system is directly above the plate.

Similarly, it can be seen that if both coils are held vertically, the field will, under certain conditions, have the appearance as in Fig. 44b, there being a maximum *augmentation* of the primary field (a positive anomaly) when the plate is midway between the transmitter and the receiver.

In Fig. 45a are shown the results of some small-scale laboratory experiments on a 1 mm thick zinc plate at different depths below the plane of the coils, these being horizontal and 100 mm apart. Fig. 45b is a simplified vector diagram on which are plotted the maximum real and imaginary component anomalies measured above several vertical conducting plates at two different depths (20 and 40 mm) and with different values for the factor ρ/vd where ρ is the resistivity of the plate, d its thickness and v the frequency.

It will be noticed that, as in Fig. 41, a single field observation suffices to determine the depth and ρ/vd for a *thin* plate and also that, for the same depth the field of all plates with the same ρ/vd is practically identical. (The resistivities of the plates used are Cu 16.5, Al 29, Zn 60 and Pb 210 nano-ohm-m).

Experiments show that if the factor ρ/vd is kept constant the thickness d must not exceed a certain value d_{max} if the field is to remain unchanged. This value depends upon ρ and v. For a given v, d_{max} is smaller the better the conductor and for a given ρ it is smaller the higher the v. The result of this limitation on d is that the vector curve for each depth splits up as

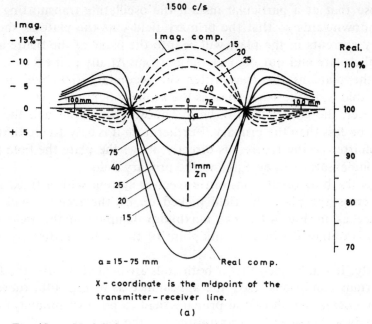

1500 c/s

Imag.

-15%
-10
-5
0
+5

Imag. comp.

15
20
25

40

100mm   0   75

1mm
Zn

75
40
25
20
15

Real.

110 %

100

90

80

70

a = 15 - 75 mm   Real comp.

X - coordinate is the midpoint of the
transmitter - receiver line.

(a)

Fig. 45a.   Model electromagnetic experiments on a Zn conductor

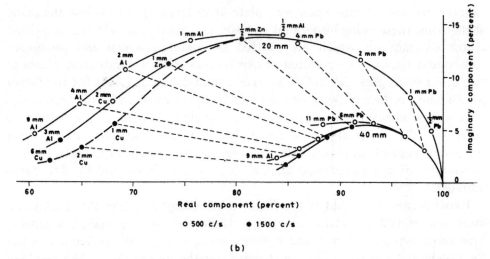

Imaginary component (percent)

-15
-10
-5
0

1 mm Al   ½ mm Zn   ½ mm Al
                      4 mm Pb
             20 mm   2 mm Pb

2 mm Al   1 mm Zn
4 mm Al   2 mm Cu   1 mm Pb
9 mm Al   ½mm Pb
3 mm Al   1 mm Cu   11 mm Pb 5mm Pb
6 mm Cu   40 mm
          2 mm Cu   9 mm Al

60   70   80   90   100

Real component (percent)

o 500 c/s   • 1500 c/s

(b)

Fig. 45b.   Vector diagram for vertical conducting plates at two depths for the system in
Fig. 44a (coil distance 100 mm)

118

shown. This makes it possible, within certain limits to estimate ρ and d separately by appropriately varying the frequency (99). This was first pointed out by Hedström (100).

Vector diagrams as in Fig. 45b are of great value in the quantitative interpretation of field surveys. Sets of them for different depths, conductor dips, lengths etc. have been published by Nair *et al* (175).

The response of *inclined* plates to the horizontal transmitter-receiver system is shown in Fig. 46a. In all these cases there is a minimum in the imaginary component (the I scale is negative upwards!) almost exactly above the top edge of the conductor. The minimum in the real component occurs at a point slightly towards the 'inside' of the plate. However, on a conductor poorer than that shown both minima are shifted towards the inside.

The R and I responses *along the line of measurements* are plotted in Fig. 46b as vector diagrams. The figure furnishes an interesting comparison of how the secondary vector 'swings' in the various cases as the coils are brought towards the conductor from infinity.

In Fig. 47 are reproduced the results of some field measurements on a thin sulphide conductor containing about 1 per cent copper as chalcopyrite, $CuFeS_2$. The conductor has a gentle dip of $15°$ with the horizontal. The measurements were made at two different frequencies and two transmitter-receiver separations. The smaller coil separation maps more detail in the anomalies but the effect of near-surface conductivity variations is liable to be accentuated. At the same time, however, the use of a lower frequency emphasizes the relatively deeper conductivity variations. By a judicious application of such measurement techniques it is possible to derive considerable detailed information about the conductors evinced by the electromagnetic anomalies obtained with the moving source-receiver system (99).

5.6 Theoretical approaches

The problem of calculating theoretically the response of a conductor to an electromagnetic field is that of solving Maxwell's equations under appropriate boundary conditions. On account of its complicated nature the treatments have so far been confined to relatively simple conductor models such as a semi-infinite earth (102), a sphere (103–105), a cylinder (105a),

(a)   (b)

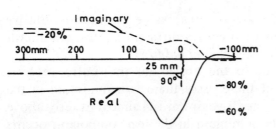

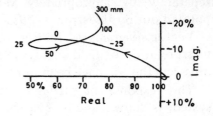

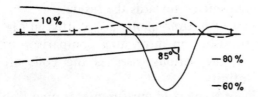

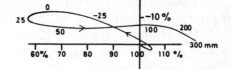

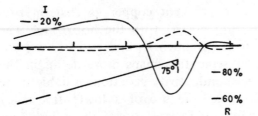

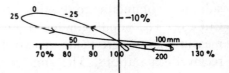

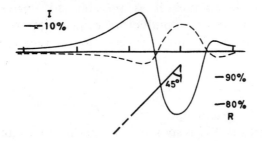

Conductor 1mm Zn
Frequency 1500 c/s
Coil separation 100 mm

Fig. 46.   *Real and imaginary component anomalies with the system in Fig. 44a across dipping sheets*

120

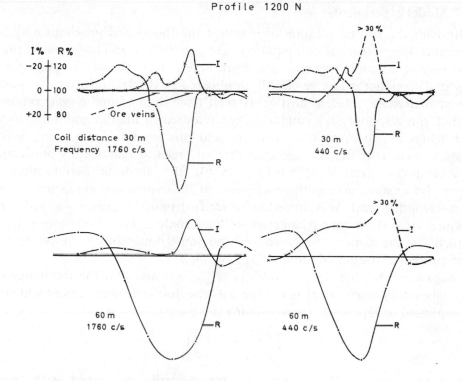

Fig. 47.   *Survey profiles with the moving source-receiver method in Fig. 44a (Courtesy Roros Copper Works, Norway)* (Hor. scale 6 mm = 30 m)

vertical and horizontal sheets (*106, 107*) and an earth whose conductivity is a function of the depth only (*108, 109*).

Common to all these treatments is the assumption that the displacement currents in the earth materials can be neglected, that is, $\rho \epsilon v \ll 1$, where ϵ is the absolute dielectric constant (farad/m = mho-s/m).

Reported values for ϵ range from about 17×10^{-12} to about 75×10^{-12} farad/m for most dry rocks but may be as high as $170-700 \times 10^{-12}$ farad/m depending upon the water content. Even with the highest ϵ-values however, the above assumption is seen to be justified with a good margin in view of Table 6, even if the frequency be as high as 100 kc/s.

5.7 Model experiments

Although the formal solution of many of the theoretical problems may be available the numerical computations are generally formidable and require recourse to a computing machine. Fortunately, however, the response of a 'natural' conductor can be exactly duplicated in the laboratory on a small, convenient scale. This is most easily seen from dimensional considerations.

Let the response of a conductor be measured in the dimensionless form S/P where S and P are the secondary and primary fields at a point. Since displacement currents are neglected the only relevant parameters producing this response are: ρ (ohmm), ν (s^{-1}), the absolute permeability μ (henry/m = ohm-s/m) and the linear scale of the experiment characterized by some length d (m). It is immediately verified that the quantity $\gamma = \rho/\mu\nu d^2$ formed from these four parameters is dimensionless. Therefore every system which has the same γ must produce the same dimensionless response S/P irrespective of the actual values of ρ, μ, ν and d.*

Suppose now that $d_{\text{full scale}}/d_{\text{laboratory}} = n$ and that the frequency of the laboratory experiment is the same as the 'full scale' experiment which it is supposed to represent. Then equating the γ's,

$$\left(\frac{\rho}{\mu}\right)_{\text{f.s.}} = n^2 \left(\frac{\rho}{\mu}\right)_{\text{lab.}} \tag{5.3}$$

Since, however, in both cases we are generally concerned with 'non-magnetic' conductors $\mu = \mu_0$, the permittivity of free space, so that

$$\rho_{\text{f.s.}} = n^2 \rho_{\text{lab}}. \tag{5.4}$$

This means that if the linear dimensions in the laboratory experiment are n times smaller than those in the full-scale work, a laboratory conductor with resistivity ρ will correspond to a geometrically similar full-scale conductor of resistivity $n^2 \rho$.

Thus suppose that 10^{-3} m (1 mm) in the laboratory is chosen to represent 1 m in full scale ($n = 10^3$). A transmitter-receiver separation of 100 m will then be represented by 100 mm, a conductor depth of 50 m by 50 mm and so on for all the other lengths involved. The response of a 1-mm thick zinc

---

* For a wider application of such considerations see the Methuen Monograph, *The Method of Dimensions*, by A. W. Porter

plate ($\rho = 6.0 \times 10^{-8}$ ohmm) will then be the same as that of a 1-m-thick full-scale conductor, e.g. an ore vein, of resistivity $6.0 \times 10^{-8} \times 10^6 = 0.060$ ohmm. Many pyrite and chalcopyrite ores have resistivities of this order (*110*).

The choice of model conductors is limited more or less to metal sheets and these do not show a very wide range of conductivities. However, since we have the scale factor, the frequency and to some extent the magnetic permeability as additional variables at our disposal, practically any conductor in full scale can be electromagnetically simulated in the laboratory.

Model experiments have already been referred to in Sections 5.3 and 5.5. It should be easy now on the basis of the above discussion to 'translate' these to full scale.

5.8 Depth penetration

The question how deep electromagnetic waves penetrate into the ground is of great importance in geophysics. If the ground were perfectly insulating the waves could penetrate to any distance. However, owing to the finite conductivity of most surface formations and the underlying rocks the incident energy is absorbed and the amplitude of the waves decreases exponentially in traversing the conductors.

It is customary in geophysical work to call the distance in which the amplitude is reduced to $1/e$ of its surface value as the *depth penetration d*. For plane waves traversing a 'non-magnetic' conductor

$$d = 503.8 \sqrt{\left(\frac{\rho}{\nu}\right)} \text{metres} \qquad (5.5)$$

which shows that the depth penetration decreases with a decrease in resistivity and an increase in frequency.

If $\rho = 2,000$ ohmm, a high value typical of relatively dry glacial moraines, sandy clays, sandstones and chalk, the depth penetration will be about 700 m at $\nu = 1,000$ c/s, 225 m at $\nu = 10,000$ c/s and only 22.5 m at $\nu = 100$ kc/s (radio frequencies). However, owing to the presence of water containing dissolved salts many surface formations have resistivities as low as 100 ohmm and d may be reduced to about 150 m even at $\nu = 1,000$ c/s. In sedimentary formations shales with resistivities of the order of 1 ohmm are quite common and in such cases frequencies as low as 10 c/s will be required to obtain a depth penetration of 150 m.

The relative response of deep-seated conductors increases and that of shallow ones decreases as the frequency is lowered. This is illustrated by Table 7 constructed with the help of Fig. 41 for the case of two horizontal conductors, each 10 m thick but having different resistivities. The use of very low frequencies for deep exploration is however limited by the fact that the absolute magnitude of the signals decreases more or less in proportion to the frequency.

TABLE 7

Relative frequency response of shallow and deep conductors

Conductor		Depth	Amplitude of secondary field (% primary field)	
No.	Resistivity (ohmm)	*m*	1,000 c/s	200 c/s
1	2.0	30	75.9	51.0
2	0.4	75	5.0	14.0

Experiments have sometimes been reported in which radio and radar frequencies penetrated several hundred metres of rock. From the above it will be clear that these rocks must have been exceptionally dry and highly resistive. Such low conductivities are however rare so that radio and radar waves are normally of no use in geophysical exploration (*111, 112*) except for the mapping of surface conductivity variations.

Apart from the depth penetration defined above there is another magnitude which is often referred to as the (practical) depth penetration. This is the maximum depth at which a conductor may lie and yet give a recognizable electromagnetic anomaly. This depth depends on the nature and magnitude of the stray anomalies ('noise') caused by near-surface conductivity variations, on the geometry of the deep conductor and on the instrumental noise. Experiments show that in the ideal case where the first noise type can be neglected the maximum practical depth penetration of the various electromagnetic arrangements is between about 1 and 5 times the separation between the transmitting and receiving systems.

5.9 Transient methods (Time-domain EM)

In recent years methods in which electromagnetic energy is supplied to the ground by transient pulses instead of by continuous waves have evoked considerable interest. A method known as 'Eltran' was tried in America (*113*) in the early 1950's, in which electric transient pulses were applied to the ground through two current electrodes while the receiver consisted of two potential electrodes. A typical value for the spread of each pair of electrodes is 300 m, the distance between the pairs being relatively large. In such cases the transmitting system is virtually an electric dipole on the surface of a semi-infinite conductor. The pulse arriving at the receiver is distorted more or less depending upon the conductivity of the ground, the distance to the receiver and the reflections it has suffered. Fig. 48a shows the distortion of a transmitted square wave pulse.

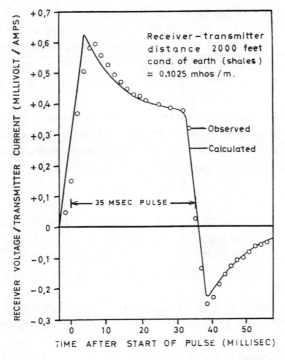

Fig. 48a.   *Distortion of a square pulse (Ref. 113)*

Of course, the energy in transient methods can also be supplied purely inductively by pulses in an insulated cable loop and the receiver can be a search coil as in the continuous wave methods. The signal in the receiver decays gradually to zero as the secondary currents in subsurface conductors dissipate on account of the electrical resistance so that the smaller the resistivity of the ground the slower is the decay of the received signal.

The transient and the continuous wave methods are related to each other through the Fourier transform (p. 199) according to which any waveform, transient or continuous, can be decomposed into a number of sine and cosine waves. Electromagnetic waves are delayed in passing through any conductor. In the case of sinusoidal waves this delay is measured as a phase shift while with pulses the actual delay-time is measured. It may be noted, for instance, that with a frequency of 1,000 c/s a phase shift of 1° is equivalent to a delay of 1/360 x 1/1,000, that is 2.78 μs. The transient signal received is a superposition of the secondary amplitudes of all the frequency components, each with its characteristic delay.

From section 5.8 it is evident that the high frequency components of a transient pulse are damped out at shallower depths than the low frequency ones. Consequently, the signals appearing at later times are governed by conductors lying at depths greater than the depths which govern the earlier signals.

On a non-homogeneous earth we can calculate an 'apparent conductivity,' c_a (mho/m), for each instant of measurement, defined as the conductivity that a homogeneous half-space must have to yield the secondary field actually observed at that instant with the given transmitter-receiver set-up. This concept of c_a *as a function of time* is due to Morrison *et al* (*176*). Fig. 48b shows c_a on a two-layer earth for a few different combinations of conductivities. We see that at early times the c_a curve represents the conductivity of the upper layer (σ_1) and at later times it approaches the conductivity (σ_2) of the lower layer (infinite substratum) asymptotically, a situation entirely analogous to the variation of ρ_a with L in section 4.4.

Transient methods possess certain advantages. It is clear from section 5.8 that either very great primary field strengths at high frequencies or, if the field strengths are moderate, very low frequencies must be used in the continuous wave methods, if subsurface conductors below highly conductive

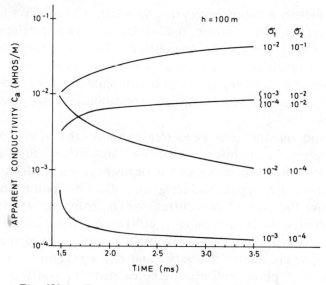

Fig. 48b.   Transient e.m. reponse of a two-layer earth

overburden are to be excited. In either case the instrumentation becomes very difficult, cumbersome and costly. The production of very high transient field strengths, on the other hand, is easier although the difficulties involved should by no means be underrated. Experiments have been reported in which a power of 700 kW was dissipated in 100 ms through a square single-turn loop of side 300 m (*177*).

Transient methods are still largely in the nascent stage but considerable theoretical work has been done on the responses of spherical and cylindrical bodies and the stratified earth (*178, 179, 180*).

5.10 Influence of magnetic permeability

We have assumed so far that the magnetic permeability of conductors is equal to that of vacuum. This however is not justified for some natural conductors, e.g. magnetite ores, for which the relative permeability may be as much as 10.

From Section 5.7 it is evident that the effect of increasing the permeability n times is the same as that of increasing the conductivity $(1/\rho)$ n times. However, besides the electromagnetic field there is in the case of magnetic conductors a purely *magnetic* secondary field due to the oscillating induced magnetism. The former field increases with frequency and is, generally speaking, opposed to the latter. There is therefore a critical frequency for a magnetic conductor above which its response is typically electromagnetic (e.g. damping in the real component) and below which it is predominantly magnetic (e.g. with an appearance as for the field of a magnetic pole).

In Turam and moving source-receiver methods the effect of magnetic permeability generally manifests itself by "distorting" the ratio or real component curves from their symmetric or near-symmetric form into curves of the antisymmetric type exhibiting an inflection point instead of an extremum above the current concentration. The converse is the case for the curves in the compensator method (Fig. 40b) on magnetic conductors. These are 'distorted' to symmetric types. In other instances anomalies which are 'normally' positive are rendered negative and vice versa (Fig. 65).

The influence of permeability must be carefully considered in fieldwork when magnetic conductors are suspected in an area (*163*).

It is worth noting however that the magnetic permeability affects mainly the amplitude of the total secondary electromagnetic field rather than its phase.

5.11 The Telluric methods and AFMAG

It has been known for a long time that electric currents of different kinds are flowing through the ground producing a potential difference between two points. Some of these currents are artificial, caused by electric railways, power lines, etc. while others may be natural, e.g. those due to the self-potentials of sulphide ores. Besides these local phenomena creating potential differences in limited regions there are currents caused by astronomic phenomena, such as the solar electron streams, the rotation of the earth, etc. which cover extremely large areas. These currents flow in vast sheets involving the entire surface of the earth and are therefore called telluric currents.

The telluric electric fields are of the order, say, of 10 mV/km and are constantly fluctuating in direction and magnitude at any point.

If X and Y are two orthogonal components of the telluric field at any point on the earth's surface they are related to the components X_0 and Y_0

simultaneously existing at a base point by the simple linear equations:

$$\left.\begin{array}{l} X = aX_0 + bY_0 \\ Y = cX_0 + dY_0 \end{array}\right\} \tag{5.6}$$

The matrix $\|ab/cd\|$ is characteristic of the electric properties of the surface down to a depth of a few kilometres at the point under consideration.

On account of the linear form of equations (5.6) the relationships between the time derivatives of the components are exactly the same as those between the components themselves. In the Telluric method the variations ΔX, ΔY, ΔX_0, ΔY_0 in successive intervals of, say, 10 s are determined from the simultaneous photographic records of the potential difference between two X and two Y electrodes at the field point and two X_0 and two Y_0 electrodes at the base.

It is easy to see that the successive normalized components $\Delta X_0/\Delta R_0$ and $\Delta Y_0/\Delta R_0$ where $\Delta R_0^2 = \Delta X_0^2 + \Delta Y_0^2$ define a circle while the vectors $\Delta X/\Delta R_0$ and $\Delta Y/\Delta R_0$ define an ellipse. The ratio of the area of the ellipse to that of the circle at the base is a convenient measure of the relative telluric disturbance at the field point.

Telluric currents obey the ordinary laws of electricity. Thus, for instance, if, E is the electric field at a point, then $E = \rho j$ (Ohm's law) where j is the associated current density. Consider a vertical contact between two media (Fig. 33a) and imagine a telluric current density in the plane of the figure. If $\rho_2 < \rho_1$, then j normal to the contact will increase as we approach the contact from medium 1 and by Ohm's law E, that is $\Delta R/\Delta R_0$, will also increase. Conversely, $\Delta R/\Delta R_0$ will fall as we approach the contact from medium 2. E must therefore be discontinuous across the contact. Actually its variation is very similar to that of ρ_a in Fig. 33a or 33b.

In general all geologic structures which tend to disturb the horizontal flow of the telluric current sheets, e.g. salt domes, folded strata, buried ridges, etc. will cause telluric anomalies.

Detailed descriptions of the telluric method will be found elsewhere in the literature (*114, 115, 116, 183*).

The magneto-telluric method.
This is in a sense a further development of the Telluric method. Briefly, it

involves a comparison of the amplitudes and phases of the electric and magnetic fields associated with the flow of telluric currents.

The measurement of the electric field is relatively easy as indicated above, that of the magnetic field is considerably more difficult since we are concerned with frequencies around 1 c/s and down to 0.001 c/s or less. It is necessary to have coils with highly permeable cores and some 20–30,000 turns of wire. Such a coil can be about 2 m long and weigh some 30–40 kg. The voltages induced in the coil are detected by very high gain, low-noise amplifiers and the entire equipment for the measurement of the magnetic field can easily weigh some 70 kg or thereabouts.

Consider a telluric current sheet of frequency v flowing in an electrically homogeneous earth of resistivity ρ. The depth penetration of such a sheet, that is, the depth at which the current density in it falls to $1/e$ of its value at the surface, is given by equation (5.5).

It can be shown that the surface electric and magnetic fields E and H respectively are horizontal and, of course, orthogonal. Their amplitudes are related by the equation:

$$\rho = \frac{0.2}{v}\left(\frac{E}{H}\right)^2 \tag{5.7}$$

where ρ is in ohm-m, E in mV/km and H in $\gamma(10^{-5}$ gauss). Their phases differ by $\pi/4$, H lagging behind E.

If, then, we measure E and H at a definite frequency the first indication of the non-homogeneity of the earth will be that the phase difference θ will not be $\pi/4$. Secondly, ρ calculated from measurements at different frequencies will not be the same. However, we can always define an *apparent* resistivity ρ_a by equation (5.7).

On determining ρ_a and θ as functions of frequency by actual measurements we obtain magneto-telluric soundings in a manner analogous to the electric drillings in Chapter 4 where, however, the current penetrates deeper because the electrode separation is increased.

Theoretical calculations of ρ_a and θ as functions of v for a horizontally stratified earth have been made by Cagniard who has also given one of the classic accounts of this method (*117*).

Fig. 49a and 49b show ρ_a and θ for a two-layer earth. The major application of this method in the future is likely to be in elucidating very

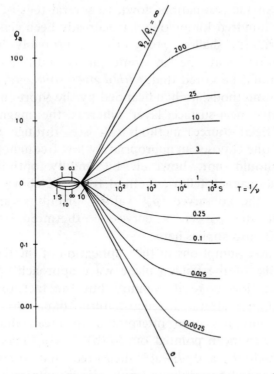

Fig. 49a. Magnetotelluric sounding: ρ_a against ν on a two-layer earth

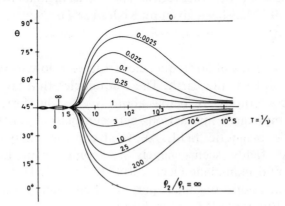

Fig. 49b. Magnetotelluric sounding: θ against ν on a two-layer earth

deep structures. In fact, soundings down to several tens of kilometres, and even a couple of hundred kilometres, have already been claimed. The utility of magnetotellurics in shallow prospecting is probably limited, but one possible application may be in some areas with highly conductive overburden. It should be noted that, *with a given source-receiver separation*, the d.c. resistivity methods are handicapped by the short-circuiting effect of a high-conductivity near-surface layer whereas the magnetotelluric (and similar, but, artificial source) methods can 'see' through the layer to the required depth if one chooses an appropriately low frequency.

The reader should not, however, be unduly optimistic about the possibilities of the magnetotelluric method from the above, because usually the scatter of the observed ρ_a values prohibits an unambiguous interpretation of the ρ_a vs. v curves. Futhermore, the principle of equivalence (p. 90) also applies here.

One of the basic assumptions in the application of the telluric methods is that of uniformity of the fields ('plane wave approach'). There has been considerable discussion as to its validity but the fact that in areas with uniform sedimentary strata, magnetotelluric data are repeatable and independent of time has been interpreted to mean that the fields are uniform. It has also been pointed out (*84*b) that ρ_a may depend on the direction in which E and H are measured and therefore a correct interpretation of sounding graphs in some cases requires a tensor analysis of the $\rho_a - v$ curves.

A good review of the theoretical and observational work in magneto-tellurics up to 1970 has been given by Keller (*184*).

AFMAG.

Natural magnetic fields of all frequencies from very low to very high ones are reaching any point on the earth. The frequencies that are exploited in the magnetotelluric methods are below about 1 c/s. The corresponding energy comes from complicated interactions between plasma emitted from the sun and the earth's magnetic field. Above 1 c/s the energy of the natural electromagnetic fields seems largely to come from local and distant thunderstorms and man-made electrical disturbances. Depending upon local and seasonal variations the main part of their energy seems to be in the region of a few hundred c/s to a few kilocycles/s.

The space between the ionosphere and the earth's surface acts as a wave guide for these fields with the result that their vertical component is normally very small. Their amplitudes and directions tend to be random or rather quasi-random. Normally a search coil at any point will show a marked horizontal plane of polarization for these waves and a diffuse azimuth in that plane. However, in the vicinity of highly conductive bodies the plane of polarization tilts out of the horizontal while the azimuth becomes more definite, and the presence of the conductor is thereby revealed. Usually the field strength shows an inflection point above the conductor and is flanked on the sides by a maximum and a minimum (cf. Fig. 49c).

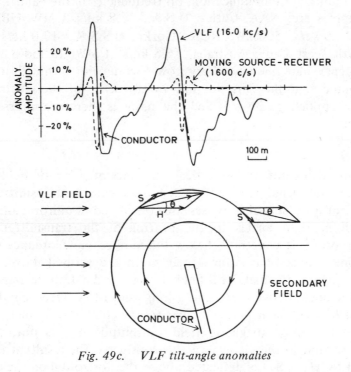

Fig. 49c.   VLF tilt-angle anomalies

The name Afmag is derived from the fact that the fields picked up by the search coil are *a*udio *f*requency *mag*netic fields. On account of the relatively high frequency the coil weight need be only a fraction of that required in the magnetotelluric method. The tilt measurements are made for two different frequencies, one high and one low, and the response ratio 'Low/High' provides a measure of the conductivity of the conductor. It will be seen from the discussion in section 5.2 that if $S\cos\phi$ or α is used as a measure of the response such a ratio is greater than 1 for good conductors and less than 1 for bad ones.

5.12 VLF methods (H-mode and E-mode)

Powerful radio transmitters situated in different countries transmit unmodulated carrier waves, either continuously or with Morse code, for purposes of military communication, on frequencies in the band 15–25 kc/s. Some examples are: NAA, Cutler, U.S.A., 17.8 kc/s, 1 MW; GBR, Rugby, England, 16.0 kc/s, 500 kW; ROR, Gorki, U.S.S.R., 17.0 kc/s, 315 kW; NWC, North West Cape, Australia, 15.5 kc/s, 1 MW. In radio technology these frequencies are known as *Very Low Frequencies*, although it should be borne in mind that they are not low in the sense of the electromagnetic methods of applied geophysics and the name is therefore improper in this connection.

H-mode.

As the transmitting antenna is vertical, the lines of magnetic field are circles in the horizontal plane. At very large distances from the transmitter the field (H_0) may, for all practical purposes, be considered as uniform and as having an azimuth at right angles to the bearing of the transmitter from the observation point. It penetrates below the surface in accordance with (5.5). The situation is exactly as in the Afmag method described above, except for the high frequency, the high VLF field strength and a definite azimuth.

If α is the angle between H_0 and a long, sheet-like, steep conductor, the component $H' = H_0 \sin\alpha$ perpendicular to the conductor induces secondary currents in it, whose magnetic field (amplitude S_0) is not, in general, horizontal so that in the vicinity of the conductor the resultant field vector (as defined by $\mathbf{H_0} + \mathbf{S_0}$) is deflected above the horizontal on the side of the transmitter and below it on the opposite side (Fig. 49c). In actual field work, not this vector but the tilt of the major axis of the polarization ellipse is determined. If $\alpha = 0$, there is no induction in the conductor. Hence the choice of the VLF station must be made with due consideration to the geological strike.

Measurements are carried out as follows. A coil (or solenoid) tuned to the frequency of the selected VLF station is held with its axis horizontal and turned so that a minimum signal is obtained. The vertical plane perpendicular to the axis in this position is the plane in which the major axis of the (magnetic) polarization ellipse lies. (Only when far from any conductor will the coil axis in this position indicate the bearing of the VLF station.) Next, the coil is turned through 90°, the axis continuing to be horizontal, and finally it is tilted around the horizontal line through its centre and lying in its plane, * until a minimum signal is obtained. In this final position the major-axis of the polarization ellipse lies in the plane of the coil. Perfect null is not obtained because the out-of-phase field represented by the minor axis is cutting the coil in its tilted position.

If ΔV is the amplitude of the vertical component of the secondary field due to a conductor, then it can be shown that† $2\Delta V \cos \phi / H = \tan 2\theta$ where H is the total horizontal field, θ is the tilt of the coil axis from the vertical (that is, of the major axis from the horizontal) and ϕ is the phase lag of the secondary field. Since usually $H \approx H_0$ and $\Delta V \ll H$ we can write $\theta \approx \Delta V \cos \phi / H_0$ so that $100 \times \theta$ (in radian) gives $\Delta V \cos \phi$ in percent of the local primary VLF field strength H_0.

It will be observed that the VLF method is essentially a tilt-angle method (section 5.2).

The upper part of Fig. 49c shows a profile across two parallel sulphide conductors, each several kilometres long, in N. Sweden. Also shown for comparison are the anomalies with the moving source-receiver method (1,600 c/s). Extensive field tests with the method have been reported in articles (*185, 186*).

E-mode.

As far back as 1937 Norton showed (*187*) that on homogeneous, conductive ground, the electric vector E of the VLF field is elliptically polarized and that, along the ground, at distances beyond about one wavelength from the antenna, the angle λ that the major axis of the ellipse makes with the vertical and the ratio minor-axis/major axis remain virtually constant at all points.

---

* If a solenoid is used this line will obviously be the line perpendicular to the length of the solenoid

† The relation is far from being intuitively obvious

The exact expression for λ is complicated. Fortunately, since $\rho\epsilon\nu \ll 1$ (p. 121) for $\nu < \sim 20\,kc/s$ in geophysical applications, it can be simplified and then gives

$$\tan \lambda = \sqrt{\pi\rho\nu\epsilon_0}$$

$$= 0.527 \times 10^{-5} \sqrt{\rho\nu}$$

where $\epsilon_0 = (1/36\pi) \times 10^{-9}$ farad/m is the dielectric constant of free space. In view of the condition imposed this equation is valid for VLF fields only if $\rho < \sim 10^4$ ohm-m.

The determination of λ will yield ρ for a homogeneous ground. On non-homogeneous ground we may define an *apparent resistivity* $\rho_a = 3.60 \times 10^{10}$ $(\tan^2 \lambda/\nu)$ or, for small values of λ,

$$\rho_a = \frac{1096 \times 10^4}{\nu} \lambda^2 \quad (\lambda \text{ in degrees}) \tag{5.8}$$

In principle, λ can be determined by orientating a long wire-antenna so that the signal in it is minimum, in which position the antenna coincides with the minor axis. Since E is almost vertical the antenna orientation for the minimum signal position is roughly horizontal. But on account of the cumbersome antenna lengths required the *E*-mode of VLF measurements is not very convenient for ground surveys. It has, however, been successfully adapted to airborne work (section 8.4).

It should be noted that the *E*-mode ρ_a variations across vertical discontinuities are qualitatively exactly the same as those in Figs. 33a and 33b. Indeed, they cannot be otherwise since we are measuring the horizontal electric field in the ground in both cases.

6 · Seismic Methods

6.1 Introduction

The seismic methods of geophysical exploration utilize the fact that elastic waves travel with different velocities in different rocks. The principle is to initiate such waves at a point and determine at a number of other points the time of arrival of the energy that is refracted or reflected by the discontinuities between different rock formations. This then enables the position of the discontinuities to be deduced.

The importance of the seismic methods lies above all in the fact that their data, if properly handled, yield an almost unique and unambiguous interpretation.

The standard method of producing seismic waves is to explode a dynamite charge in a hole. Attempts have been made to obtain the seismic energy by other means, e.g. weight-dropping, electrodynamic shaking, etc. but the energies thus produced are not generally sufficient and the methods have found only a limited application.

In recent years, however, there has been a general tendency towards replacing explosive sources by other means. We can classify the present-day seismic sources as follows: (1) Solid chemical explosives (Dynamite, Aquaseis, Flexotir, Primacord, Maxipulse are some of the trade names), (2) Compressed air sources (PAR, Seismojet, Terrapak etc.), (3) Electrical energy sources depending for their action on the sudden movement of a piston or a plate by transducer devices (Boomer, Pinger, Sono Probe, Sparkaaray etc.), (4) Gas exploders (Acquapulse, Dinoseis, Deltapulse etc.), (5) Mechanical Impulse sources (Hammer, Thumper), (6) Implosive sources (e.g. Hydroseis) and (7) Vibratory sources (Vibroseis) (*188, 189*).

Of these, Vibroseis falls in a class by itself on account of its continuous input signal. It is treated separately later on in section 6.12. The others produce 'conventional' sudden pulses of short duration and the choice depends only on the operational requirements of the survey.

6.2 Elastic constants and elastic waves

The basis of the seismic methods is the theory of elasticity. The elastic properties of substances are characterized by elastic moduli or constants which specify the relation between the *stress* and the *strain*. A stress is measured as force per unit area. It is compressive (or tensile) if it acts perpendicular to the area and shearing if it acts parallel to it. A system of compressive stresses changes the volume but not the shape of a body, one of shearing stresses changes the shape but not the volume.

The strains in a body are deformations which produce restoring forces opposed to the stresses. Tensile and compressive stresses give rise to longitudinal and volume strains which are measured as the change in length per unit length or change in volume per unit volume. Shearing strains are measured as angles of deformation. It is usually assumed that the strains are small and reversible, that is, a body resumes its original shape and size when the stresses are relieved.

Hooke's law states that the stress is proportional to the strain, the constant of proportionality being known as the elastic modulus or constant. The law is not strictly true and more generally stress—strain relationships have also been introduced in applied seismology, notably by Ricker (*119, 120*); yet, Hooke's law carries us a long way in the theory of elasticity.

The two moduli of immediate interest for the study of the elastic waves in the earth are the bulk modulus (k) and the shear modulus (n). Their values for some rocks and for a few common substances will be found in Table 8.

TABLE 8

Elastic constants

Substance	Bulk modulus (newton/m^2 x 10^{-10})	Shear modulus (newton/m^2 x 10^{-10})
Marbles and limestones	3.7—5.7	2.1—3.0
Granites	2.7—3.3	1.5—2.4
Sudbury diabase	7.3	3.7
Ohio sandstone	1.25	0.61
Iron (wrought)	16.0	7.7
,, (cast)	9.5	5.0
Glass (crown)	5.0	2.5
Quartz (fibre)	1.5	3.0

Elastic waves.

If the stress applied to an elastic medium is released suddenly the condition of strain propagates within the medium as an elastic wave. There are several kinds of elastic waves:

(1) In the longitudinal, compressional or P waves the motion of the medium is in the same direction as the direction of wave propagation. These are, in other words, ordinary sound waves. Their velocity is given by

$$V_l = \sqrt{\left(\frac{k + (4/3)n}{\rho}\right)} \qquad (6.1)$$

where ρ is the density of the medium.

(2) In the transverse, shear or S waves the particles of the medium move at right angles to the direction of wave propagation and the velocity is given by

$$V_t = \sqrt{\left(\frac{n}{\rho}\right)} \qquad (6.2)$$

It is evident that $V_t < V_l$.

Shear waves can be polarized in which case the particles oscillate along a definite line perpendicular to the direction of wave propagation.

(3) If a medium has a free surface there are also surface waves in addition to the above two which are 'body waves'. In the Rayleigh waves the particles describe ellipses in the vertical plane that contains the direction of propagation. At the surface the motion of the particles is retrograde with respect to that of the waves. The velocity of Rayleigh waves is about $0.9\ V_t$.

(4) Another type of surface waves are the Love waves. These are observed when the velocity in the top layer of a medium is less than that in the substratum. The particles oscillate transversely to the direction of the wave and in a plane parallel to the surface. The Love waves are thus essentially shear waves. Their velocity for short wavelengths is equal to V_t in the upper layer and for long wavelengths V_t in the substratum.

The spectrum of the body waves in the earth extends from about 15 c/s to about 100 c/s; the surface waves have frequencies lower than about 15 c/s.

In applied seismology only the P waves are of importance. In principle, however, S waves could also be used but the difficulty is to get S waves of sufficient energy. Explosions, which are the common means of generating powerful elastic waves, produce predominantly, if not exclusively, P waves.

These are converted in part to S waves on oblique reflection and although some work has been reported with such S waves they have not been put to any great use in applied geophysics.

Surface waves are incapable of giving information about structures at depth and little interest is attached to them in exploration geophysics.

Velocities.

Typical values for the velocity of P and S waves in some rocks are given in Table 9. The velocities are generally found to be greater in igneous and crystalline rocks than in sedimentary ones. In the sedimentary rocks they tend to increase with depth of burial and geologic age. Many empirical attempts have been made to represent this increase. For shales and sands Faust (*121*) finds that

$$V = 46.5(ZT)^{1/6} \text{ m/s} \qquad (6.3)$$

where Z is the depth in metres and T the age in years.

Seismic velocities can be measured in the field as well as in the laboratory on samples of rocks. Several methods for laboratory determinations using magnetostrictive pulses, ultrasonic pulses, resonances etc. have been developed (*122, 123, 124*). Well-velocity surveys are also a common method of obtaining information on velocities. A seismometer (Section 3) is lowered into a borehole, a shot is fired close to the surface and the travel time to the seismometer is noted. The seismometer is then lifted a short distance and another travel time is measured. From the difference in these times the average velocity in the material between the two positions of the seismometer can be calculated.

TABLE 9

Elastic velocities (m/s)

	Compressional	Shear
Air	330	–
Sand	300–800	100–500
Water	1,450	–
Glacial moraine	1,500–2,700	900–1,300
Limestones and dolomites	3,500–6,500	1,800–3,800
Rock salt	4,000–5,500	2,000–3,200
Granites and other deep-seated rocks	4,600–7,000	2,500–4,000

Seismic velocities often show anisotropy in stratified media; the velocity parallel to the strata is generally greater than that normal to them by an amount of the order of 10–15 per cent.

6.3 Instruments and field procedure

In the early days seismic waves were detected by a mechanical seismometer which consisted, in principle, of a heavy mass suspended as a pendulum. The mass remained stationary on account of its inertia while the suspension frame moved with the earth. Such mechanical devices have now been completely replaced in applied geophysical exploration, by small, lightweight electric detectors or *geophones*.

The simplest and the most common type is the electromagnetic geophone. A coil attached to a frame is placed between the poles of a magnet which, in turn, is suspended from the frame by leaf-springs. The magnet acts as an inertial element while the coil moves with the earth. The relative motion of the coil and the magnet produces an e.m.f. proportional to the *velocity* of the earth's motion. In underwater detection the geophones used produce a signal proportional to the excess *pressure* on a diaphragm. There is a very large variety of geophones on the market for land as well as underwater use, some of them weighing no more than 50 g.

The natural frequencies of electromagnetic geophones in reflection seismic work are usually 20 c/s or higher but may be as low as 5 or 2 c/s in refraction work. These frequencies are damped electromagnetically or by using oil, the damping being critical or slightly less than critical.

The voltage from the geophone is fed into an amplifier, usually over a transformer for reasons of impedance matching, and is then passed on to a galvanometer. Most amplifiers are provided with automatic gain control and a number of high-pass and low-pass filters which allow a selection of the frequencies to be recorded.

The galvanometer deflections are registered by means of a mirrorlight source system on a continuously running photographic film. A single film may carry signal traces from as many as 48 geophones. The record is called a *seismogram*.

Since each geophone needs an amplifier and a galvanometer of its own, compactness is of utmost importance in the design of seismic recording equipment. Indeed, the galvanometers in seismic cameras are only a few millimetres in diameter and a few centimetres in length while the use of transistors has led to amplifiers with the approximate dimensions 15 x 10 x 5 cm.

Vertical time lines, generally running across the entire width of the film, are superposed on the record. These are generated by a light beam which passes intermittently through a slot in a rotating disc whose speed is accurately controlled by a suitable device such as a tuning-fork driven oscillator. Events on a seismogram can be read to a millisecond.

With the development of magnetic tape recording and automatic processing (section 6.10) optical recording may now be said to be largely obsolete except for refraction seismic work. However, the final displays of reflection seismograms for interpretation are always visual ones obtained after the processing of tape data.

Field procedure.
When the area of investigations has been decided upon the first step in a seismic survey is to select locations for shot-holes. If only shallow holes are needed they can be prepared by augering but generally drilling by machine is necessary in deep exploration since the shot-holes may be as much as a hundred or more metres deep. The depth of holes and the weight of dynamite for the shot (a few grammes to several kilogrammes) are important factors controlling the quality of the seismogram record (*125*).

The next step is to plant the geophones firmly on the ground which may sometimes entail burying them below the surface. In most of the work the geophones are placed along a straight line (called a profile) through the point on the surface vertically above the shot (epicentre). This procedure is known as profile shooting but other shooting patterns adapted to particular problems are also used (*14, 16, 62*).

The geophones are connected to the recording equipment by long cables, the amplifier gains and filters are appropriately set and when the recording film is set in motion the shot is fired. The film is stopped after a few seconds when the ground motion has substantially subsided.

The moment of the shot is registered on the photographic film as a break in a continuous trace along one edge of the film.

6.4 The refraction method
The basis of the refraction method is the extension of Snell's law in optics to seismic waves. If a layer in which the waves have a velocity V_1 is underlain

by another layer 'with velocity' V_2 then by Snell's law

$$V_1/V_2 = \sin i_1/\sin i_2 \qquad (6.4)$$

where i_1 and i_2 are the angles of incidence and refraction for the seismic ray (Fig. 50a). The ray SA with the *critical angle* of incidence $i_c = \sin^{-1} V_1/V_2$ is refracted so that $i_2 = 90°$ and travels along the boundary between the two media. Obviously, this is possible only if $V_2 > V_1$.

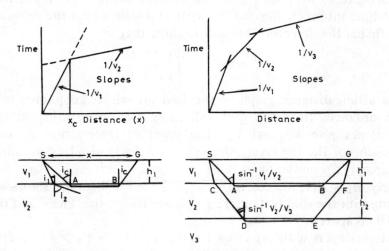

Fig. 50. (a, left; b, right) Time–distance graphs

Due to this ray the interface is subjected to oscillatory stress and each point on it sends out secondary waves and rays such as BG emerge in the top layer at the angle i_c to reach the geophone G.

If G is near to the shot S (which is assumed to be on the surface) the first 'kick' of the galvanometer will be due to the arrival of the direct wave along SG. After some time a second kick will be observed corresponding to the arrival of the refracted wave. However, if SG is sufficiently great the first arrival will correspond to the wave SABG which will have overtaken the direct wave because of a higher velocity along the path AB.

The travel time for the direct wave is $t = x/V_1$ and its plot against x will obviously be a straight line through the origin, that is the coordinate of the shotpoint, with the slope $1/V_1$ (Fig. 50a). Remembering that $\sin i_c =$

V_1/V_2 it is easily proved that the travel time equation for the ray SABG is

$$t = \frac{x}{V_2} + \frac{2h_1(V_2^2 - V_1^2)^{1/2}}{V_1 V_2} \tag{6.5}$$

which is also a straight line but with a slope $1/V_2$ and an intercept on the
t-axis given by the second term in (6.5). This term is also called the
delay-time (t_1).

Equating (6.5) with x/V_1 gives the distance-coordinate at which the two
straight lines intersect. Beyond this critical distance (x_c) the refracted wave
arrives first at the detector. It is easy to show that

$$x_c = 2h_1 \sqrt{\left(\frac{V_2 + V_1}{V_2 - V_1}\right)} \tag{6.6}$$

Thus a time-distance graph of the first arrivals at geophones planted at
various distances from the shot will show two intersecting straight lines
whose slopes give V_1 and V_2. The point of intersection x_c yields the
thickness h_1 of the top layer. Alternatively h_1 may also be determined from
the intercept (t_1) on the t-axis (6.5).

Clearly, $h_1 = V_1 t_1/(2 \cos i_c) = V_2 t_1/(2 \cot i_c)$. Fig. 50c shows a typical
recording with the shot instant, 12 geophone traces, the 'kicks' and the times
lines (10 ms apart).

For three layers with velocities V_1, V_2 V_3 $(V_3 > V_2 > V_1)$ there will be
two critically refracted rays, SABG along the first interface and SCDEFG
along the second one (Fig. 50b). As before, at short distances the direct ray
SG will arrive first. As the distance increases the ray SABG will overtake SG
and arrive first while at still greater distances the first kick of the
galvanometer will signify the arrival of the ray SCDEFG. The time-distance
plot then consists of three intersecting straight segments with slopes $1/V_1$,
$1/V_2$ and $1/V_3$. The thickness h_1 of the top layer is calculated from (6.6)
and it can be shown that the thickness of the second layer is given by

$$x'(V_3 - V_2) = \frac{2h_1}{V_1}[V_2\sqrt{(V_3^2 - V_1^2)} - V_3\sqrt{(V_2^2 - V_1^2)}]$$

$$+ 2h_2\sqrt{(V_3^2 - V_2^2)} \tag{6.7}$$

where x' is the second intersection point.

In terms of the intercepts t_1, t_2 of the second and third segments we have the following formulas:

$$h_1 = V_1 t_1 / 2 \cos i_1$$

$$h_2 = \frac{V_2 \{t_2 - t_1(1 + \cos 2 i_3)/(2 \cos i_3 \cos i_1)\}}{2 \cos i_2}$$

where $i_1 = \sin^{-1}(V_1/V_2)$, $i_2 = \sin^{-1}(V_2/V_3)$ and $i_3 = \sin^{-1}(V_1/V_3)$.

These relations can be extended to any number of horizontal layers, one below the other, provided the velocity in each layer is greater than that in the layer immediately above. In general there will then be as many distinct segments on the time-distance curve as there are layers. If, however, a layer is not sufficiently thick or does not have a sufficient velocity contrast with the adjacent layers, the segment corresponding to it may be missing on the time-distance graph of the *first* arrivals. This obviously introduces errors in the depths to the lower interfaces. Such thin layers can however be detected sometimes by recording the later arrivals.

If the velocity increases continuously with the depth, for instance approximately linearly as is often the case, the time-distance graph will be a smooth curve concave towards the x-axis.

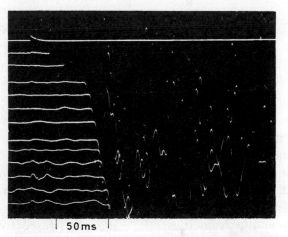

50 ms

Fig. 50c.   Refraction seismic traces showing kicks due to first arrivals

If a layer has a *lower* velocity than the one on top of it there cannot be any critically refracted ray because i_2 will always be less than i_1 in Fig. 50a. No energy can be transported along such an interface and no segment corresponding to the lower layer will appear in the time-distance curve. It is worth noting, however, that such layers *could* be detected if account is also taken of shear waves (*126*).

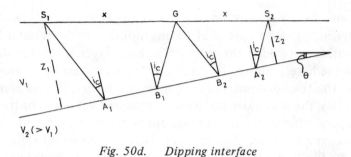

Fig. 50d.   *Dipping interface*

In Fig. 50d is shown a case where the interface between two layers is dipping at an angle with the horizontal. It is readily shown that an 'up-dip' ray such as $S_1 A_1 B_1 G$ originating at S_1 takes a time

$$t = \frac{2z_1 \cos i_c}{V_1} + \frac{x}{V_1} \sin (i_c - \theta) \tag{6.8}$$

to arrive at the geophone while for a 'down-dip' ray $S_2 A_2 B_2 G$ from a shot at S_2,

$$t = \frac{2z_2 \cos i_c}{V_1} + \frac{x}{V_1} \sin (\theta + i_c) \tag{6.9}$$

These relations, it may be noted in passing, will be interchanged if the surface is sloping and the interface horizontal.

The time-distance curve for the direct ray has the slope $1/V_1$ whether the ray comes from the shot S_1 or shot S_2. However, the segment corresponding to the refracted ray has a slope $\sin (i_c - \theta)/V_1$ when shooting up-dip but $\sin (\theta + i_c)/V_1$ when shooting down-dip. The reciprocals of these slopes are the up-dip and down-dip velocities, V_u and V_d respectively, and it is immediately seen that

$$\theta = \tfrac{1}{2}(\sin^{-1} V_1/V_d - \sin^{-1} V_1/V_u) \tag{6.10}$$

and

$$i_c = \tfrac{1}{2}(\sin^{-1} V_1/V_d + \sin^{-1} V_1/V_u) \tag{6.11}$$

(Both V_u and V_d are reckoned positive.)

The dip is directly determined from (6.10) while the depths z_1 and z_2 which complete the information about the configuration in Fig. 50d are obtained from the intercepts of the lines represented by (6.8) and (6.9) on the t-axis after substituting for i_c from (6.11).

The up-dip velocity $V_u = V_1/\sin(i_c - \theta)$ is positive if $\theta < i_c$ and negative if $\theta > i_c$. If $\theta = i_c$ it becomes infinite so that the corresponding time–distance segment is horizontal. Of course, the infinite velocity is only apparent; no energy is transmitted at this velocity.

The case of an arbitrary number of dipping, but not necessarily parallel, interfaces has been dealt with in an interesting manner by Mota (*190*). Nomograms for the rapid determination of interface depths and dips have been published by several authors (*191, 192*).

The refraction method was much used before 1930 for oil prospecting but has now been largely replaced by the reflection method. In recent years, however, refraction work has found increasing use in civil engineering projects for bedrock investigations in connection with dam-sites and hydro-electric power stations, in place of conventional drilling.

An example of a refraction profile for a hydro-electric project in Northern Sweden is shown in Fig. 51 (*127*). Each dot or cross represents the travel time to a geophone placed vertically below it on the ground surface.

The depths sounded on this profile are relatively shallow but the results nevertheless illustrate a number of points discussed above.

The steep initial segments near each shot point correspond to the upper layers and at large distances from the shots these are replaced by less steep segments. This is most clearly indicated by the left-hand geophone set-up from shot 4. The break-points between segments can be determined after appropriately smoothing the irregularities (such as I) caused by local inhomogeneities in the superficial layer.

The apparent velocity in the bedrock when the geophones occupy positions around G is less when the shot is fired from S_1 (steeper time–distance segment) than when it is fired from S_6 (almost horizontal

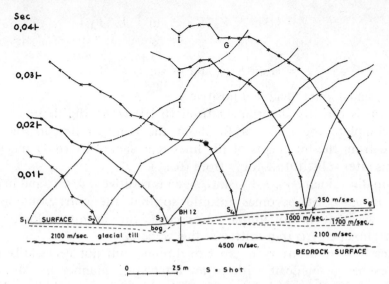

Fig. 51.   *Time–distance graphs and interpretation in a shallow refraction survey*
(Ref. 127)

segment). Evidently the ground surface and the bedrock are not parallel to each other in this region.

In the right-hand part of the figure the time–distance curves are more or less smooth indicating a fairly continuous downward increase in velocity.

6.5 The reflection method

The depth to an interface between two rock formations can also be determined by measuring the travel time of a seismic wave generated at the surface and reflected back from the interface. The energy of P as well as S waves is reflected partly as P and partly as S waves. If the reflected and the incident waves are of the same kind (both P or both S) the ordinary law of reflection applies, namely, angle of incidence = angle of reflection.

It is generally assumed that the observed reflections are only P–P reflections. This assumption is justified as a rule since near the shot point most of the explosion energy is transmitted as P waves.

In the reflection method the mutual separation of the geophones is small (e.g. 25 m) compared with the depths (as much as 5,000 m) to be sounded.

The geophones are generally spread symmetrically on either side of the shot along a straight line through it and the maximum shot-detector distance is of the order of or smaller than the depth to the shallowest horizon of interest. This arrangement ensures that the observed galvanometer kicks are due to the arrival of reflected and not refracted rays.

It is easily deduced from Fig. 52a that if V is the uniform velocity above the reflecting horizon the reflected wave arrives at G after a time

$$t = \frac{2}{V}\sqrt{(h^2 + x^2/4)} \tag{6.12}$$

so that

$$h = \frac{1}{2}\sqrt{(V^2 t^2 - x^2)} \tag{6.13}$$

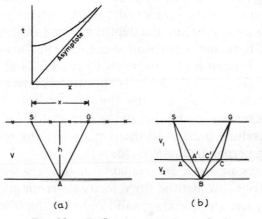

Fig. 52. *Reflection seismic method*

Either equation shows that the $x-t$ curve is a hyperbola convex towards the x-axis and with the t-axis as the axis of symmetry. The line through $x = 0$, $t = 0$ with a slope $1/V$ is its asymptote. This interesting relation is, however, of limited use in practice since the only portion of the hyperbola that is normally observed is the approximately horizontal one near $x = 0$.

If there are two reflecting horizons which separate layers with different velocities it is generally the practice to disregard the refraction of rays. A ray such as SABCG (Fig. 52b) is replaced by the ray SA'BC'G. This is justified inasmuch as the rays may be considered to be almost vertical on account of the small shot-detector separation.

The downward travel time of a strictly vertical ray is

$$t = h_1/V_1 + h_2/V_2 = (h_1 + h_2)/\overline{V} \tag{6.14}$$

where $\overline{V}$ is defined as the average velocity. If there are several layers the average velocity of the wave reflected from the nth horizon is

$$\overline{V} = \sum_1^n h_j / \sum_1^n (h_j/V_j) \tag{6.15}$$

and this may be substituted for V in equation (6.13) to obtain the depth to this horizon.

Corrections for refraction can be applied if desired. Such corrections have been calculated by Krey (*128, 129*), Lorenz (*130*) and Baumgarte (*131*).

Several procedures can be employed for the determination of $\overline{V}$ which must, of course be known before the depths can be calculated.

One procedure is to use refraction shooting. This however requires long refraction profiles if deep horizons are to be reached and adds considerably to the cost of the geophysical survey. It appears moreover that the velocities obtained do not always agree with the known velocities in reflection seismics. The difference may be due to the fact that the refraction velocities refer to waves travelling parallel to the strata while the reflection ones refer to waves travelling perpendicular thereto.

If deep boreholes or wells are available in the area well-logging (section 6.9) would be an obvious method for velocity determinations. The velocities thus obtained are, strictly speaking, valid only for the immediate vicinity of the well.

A third method employs equation (6.12). If the geophones are well separated it is possible to determine accurately the time differences ('step-outs') between the arrivals of reflections of one and the same order at each of the geophones.

If the arrival times are plotted against the squares of the geophone distances from the shot the points will lie on a straight line with a slope $1/V^2$ where V is the average velocity down to the interface from which the reflection has come. Particularly reliable results are obtained with this method if the topographic relief is minimum, the weathering in the top layer is uniform and the strata are substantially horizontal, at least over the distance occupied by the geophone spread.

The seismic record (whether on a photographic paper or on a magnetic tape) consists of wiggles. Part of a typical reflection seismogram recorded on a photographic paper is shown in Fig. 53a. A wiggle on a single trace may represent a reflected pulse or ground movement due to noise and there would be no means of distinguishing a signal if only one trace were recorded. However, noise pulses are unlikely to be exactly in phase at all the geophones, whereas a pulse reflected from a lithologic interface arrives at the various geophones approximately at the same time as the geophones are close to each other. Therefore, the reflection signals are recognized by the lining-up of the crests and troughs on the adjacent traces across the entire record. The lining-up is almost straight for the later reflections, but for the earlier reflections the pulses may lie on a slightly curved line which is actually the hyperbola represented by (6.13). As a first step in the interpretation, the step-outs or, as they are also called, the normal moveouts (n.m.o.s) of the pulses, are corrected from a knowledge of $\overline{V}$. If after this

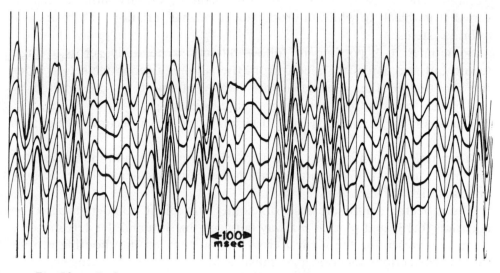

Fig. 53a.   *Reflection seismogram (British Petroleum Company Ltd., London)*

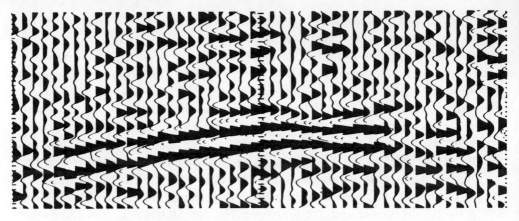

Fig. 53b.   Variable area (VAR) display

correction the pulses continue to show step-outs in the arrival times the reason can be sought in a dipping reflecting horizon.

Fig. 53b shows another display of a seismic record. Here one side of the wiggle trace is blacked-in ('variable-area' display) resulting in considerable clarity and ease in picking out the reflections. Another visual display is the 'variable-density' type in which the photographic density along a trace is varied from black to white in proportion to signal variations. There are also hybrid systems.

Sometimes it is possible to follow reflections not only across a single record but also across the record from the adjoining geophone set-up along the shooting profile. A variable-area display of this type, constructed by juxtaposing records from consecutive geophone set-ups, is shown in Fig. 53c. The scale on the vertical axis in such displays is often in seconds rather than in depth.

When reflections have been identified on a seismogram and the values of $\overline{V}$ are known the rest of the interpretation according to the ray theory is more or less straightforward.

The reflector depths are read off some convenient nomogram based on equation (6.13) and plotted midway between the shot and the detector on a diagram representing the vertical cross-section through the shooting profile. Certain corrections (Section 6.6) must however be applied to the observed travel times.

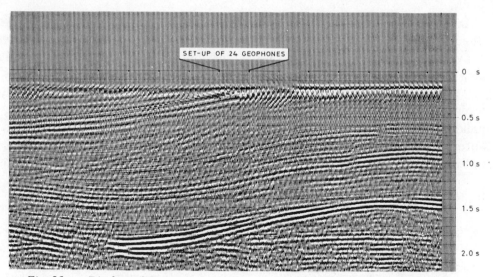

Fig. 53c. Display of juxtaposed VAR records from consecutive geophone setups (Seismograph Service Ltd., Holwood, England, and Amoco/Gas Council North Sea Group).

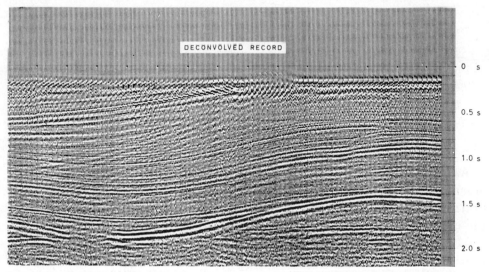

Fig. 53d. Section of Fig. 53c after deconvolution (Seismograph Service Ltd., Holwood, England, and Amoco/Gas Council North Sea Group).

153

In another method the reflection times scaled off the seismogram are directly plotted on the cross-section map. The depths are obtained by converting the times by means of the distribution of V applicable to the area.

The process of assigning depths to reflecting horizons or to selected 'events' on the basis of known or presumed $\overline{V}$ values is called *migration*. The picking-out as well as migration of events can be largely carried out automatically (*193*). Fig. 53e shows the geologic interpretation of the data in Fig. 53c. It traces the course of bedding planes and shows parts of two salt pillows between about 1.4 and 1.8s. The peripheral sink between the pillows is also apparent. Faults as indicated by the seismic sections are also marked.

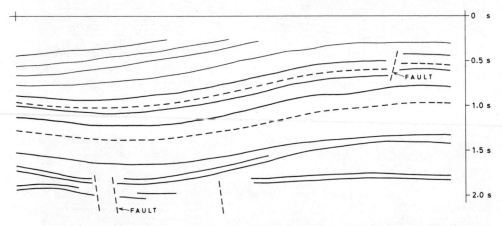

Fig. 53e.   *Geologic interpretation (by Dr. A. A. Fitch) of the section in Fig. 53c (Seismograph Service Ltd., Holwood, England, and Amoco/Gas Council North Sea Group).*

A graphical method known as the 'image method' is often used to determine the position of a dipping horizon. It can, of course, be applied to horizontal strata as well. In Fig. 54, SAG_1 is an actual ray path and AM is the extension of $G_1 A$ equal to SA. By elementary geometry we see that M is the mirror image of S in the reflecting horizon. If t_1, t_2, . . . are the travel times along SAG_1, SBG_2, . . ., etc. the arcs of radii Vt_1, Vt_2, . . ., etc. with the geophones as centres will intersect each other in M. Obviously the perpendicular bisector of SM represents the reflecting horizon.

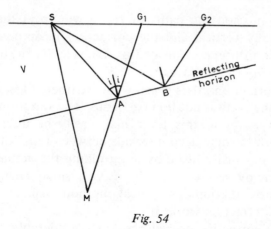

Fig. 54

6.6 Corrections to arrival times

It is usually necessary to apply two corrections to the observed travel times of seismic waves, one for the elevation differences and the other for weathering.

The theory assumes that the shot and the detector are on the same level, but in general, their elevations differ. It is customary to reduce the arrival times by projecting the shot and detector onto a common horizontal datum plane. In refraction the actual path is then replaced by the dotted path in Fig. 55. The time difference between these paths is $(SC/V_1 - AA'/V_2)$ and is to be subtracted from the observed travel time. If the elevation of the shot is h above the datum plane $SC = h/\cos i_c$ and $AA' = S'C = h \tan i_c$. A corresponding time difference originating at the detector end must also be subtracted.

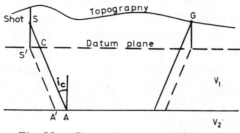

Fig. 55. Corrections to arrival times

In the reflection method the paths of the waves from the shot to the detector are practically vertical and the correction is simply equal to the shot-detector elevation difference divided by the velocity or, to be accurate, the near-surface velocity of the waves.

The weathering correction arises because the weathered low-velocity layer near the surface of the earth is not homogeneous. The variations in the delay suffered by seismic waves in this layer may easily be interpreted as the variations in the depth to some deep interface between rock formations.

The corrections are usually made by determining the actual thickness of the weathered layer by a series of shots with small charges and close shot-detector spacings but rapid graphical methods can also be devised, especially in shallow refraction work.

The details of the correction procedures vary considerably. Some of them have been described by Dobrin (*16*) and Menzel (*126*).

A theoretical treatment of the effect of the weathered layer will be found in a paper by Menzel and Rosenbach (*132*).

6.7 The seismic pulse

The simple ray theory outlined above is adequate for most purposes in applied seismology but it does not give a complete physical description of seismic phenomena. In recent years, however, a great body of literature has grown around the physical theory of the propagation of elastic waves in stratified media and, following the lead of Ricker (*133*) particular attention has been focused on the actual seismic impulse initiated by an explosion (*134*).

When a charge is detonated the material around it is permanently distorted, a cavity is formed and an outward-travelling seismic impulse (wavelet) originates. Even at distances as large as a couple of metres from the charge the maximum stress experienced by the material may exceed the range of reversible stress-strain relationships and a permanent 'set' may be evident. The distance around the shot at which the maximum stress falls just within such relationship defines the *equivalent cavity*. This region rather than the shot itself is often regarded as the source of the seismic pulse.

According to Ricker's theory, the centre of the seismic pulse travels with a velocity characteristic of the medium, and the breadth of the pulse (measured in milliseconds) increases with increasing propagation time according to some definite law. In shales, for example, the pulse breadth is proportional to the square root of the time.

If the pulse encounters a boundary between two geologic formations with different acoustic impedances (product of the velocity of elastic waves and the density of a formation) a reflection will occur and a broadened pulse will be received at the earth's surface after some time. This is illustrated qualitatively in Fig. 56a, after Anstey (*125*).

The form and breadth of the pulse and hence its frequency spectrum will depend on the travel time. The pulse will also be distorted further to a greater or less extent by the recording equipment.

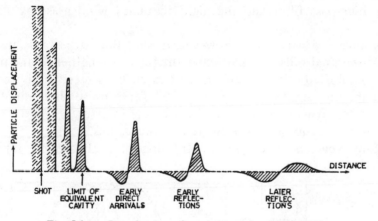

Fig. 56a. Broadening of a seismic pulse (Ref. 125)

In addition to some large velocity contrasts which cause strong reflections there will, in general, be a great number of minor velocity contrasts in a geological column. Each of these as well as other inhomogeneities will initiate a reflected pulse which arrives at the surface in broadened form. The seismogram record between two strong reflections consists not merely of 'noise' but, to a large extent, of such reflections superimposed on and interfering with each other. What remains after these *signals* are subtracted would be the noise due to surface and transverse waves, multiply reflected waves, transmissions through the weathered layer and wind.

It will now be appreciated that the seismic record contains a great deal of information about subsurface inhomogeneities but that only a small part of it can be extracted by the simple ray theory. To derive the additional information it would be necessary to have a more detailed knowledge than we at present have of the laws governing the propagation of seismic pulses in the earth.

Filtering.

The object of filtering is to exclude the noise referred to above. This noise, which is mainly due to surface waves ('ground roll') and wind, is liable to mask the reflection signals.

The common practice in filtering is to cut off the low frequency ground roll by a high-pass filter and the high frequency wind noise by a low-pass filter.

Filters have a disadvantage, however, in that they necessarily distort the shape of the signal pulse. In particular the pulse is lengthened and moreover its phase is shifted, that is, characteristics such as crests and troughs are displaced in time as illustrated in Fig. 56b for a pulse that initially has a shape of a step. The result is that filters create interference patterns on the seismogram in addition to those created by the signals.

In recent years a new kind of filtering of a theoretical rather than

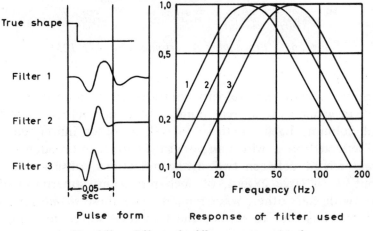

Pulse form   Response of filter used

Fig. 56b.   *Effect of a filler on a step signal*

instrumental nature has appeared. In this the seismic records are 'operated upon' so as 'to reverse the effect of the transmission medium and contract and narrow the wavelet to make up for the broadening it underwent in transmission.' The object is to derive the original shape. This procedure, known as deconvolution, requires high-speed computers. Its principle is described in section 6.11.

The subject of the old as well as the new type of filtering has been comprehensively discussed by Smith (*135*).

Multiple geophone arrays.
Besides filtering, another method is also used to eliminate or minimize the noise in seismic records. The geophones are laid out along the profile with very close separations (e.g. 10 m) and the outputs of several adjacent geophones are added together and recorded as a single trace.

Consider a surface wave of some definite wavelength. If the geometry of the geophone groups is appropriately selected the instantaneous motion of the ground due to the vertical component of this surface wave will be upwards at some geophones and downwards at others so that the sum of the outputs will eliminate the surface wave from the record.

Actually, the effect of multiple arrays is more complicated than the brief description above would indicate. It has been studied theoretically by Smith (*136*).

Attenuation.
Seismic waves are reduced in amplitude as they are propagated through the earth due to three factors: geometrical divergence, partial transmission and reflection at acoustic boundaries and absorption of energy in the medium of transmission.

The influence of the first factor is well known and can be allowed for, e.g. at large distances from the source spherical waves reduce in amplitude in inverse proportion to the distance travelled.

The transmission and reflection coefficients of a geological contact are functions of the elastic contrast between the layers in contact. In principle, a comparison of the amplitudes of the reflected pulses should provide some information about such contrasts. Although the theory of this topic is well developed (*137, 138*) the widespread use of automatic gain control in

seismic amplifiers virtually prohibits this comparison. In some magnetic tape recording seismic equipments, however, the gain of each a.g.c. channel is also recorded and it is then possible to recover the true amplitudes during
· playback.

The attenuation of plane waves due to absorption of energy is of the familiar exponential form: $A = A_0 \exp(-\delta f x / V)$ where δ is the logarithmic decrement, f the frequency, x the distance and V the velocity. Typical values of δ for the earth materials in bulk would be around $0.02-0.03$.

Two absorption mechanisms, viscosity and solid friction, have been suggested. Both mechanisms are observed in rocks but it seems that solid friction generally predominates. The presence of water apparently increases the decrement and at the same time leads to a predominance of viscous damping.

Work on the attenuation of seismic waves has been reported by Born (*139*), Collins and Lee (*140*) and Peselnick and Zietz (*141*).

6.8 Synthetic seismograms and convolution

Seismic reflections arise from changes in the acoustic impedance of the ground. If the reflection coefficient $(V_2 d_2 \quad V_1 d_1)/(V_2 d_2 + V_1 d_1)$ where d is the density, were known at each boundary it would be possible to calculate the amplitude of a reflected pulse and by combining the various pulses in the correct time relation we could 'synthesize' a seismogram record.

If the coefficient is positive, an impinging compression in a pulse will be reflected as a compression and a dilatation as a dilatation while a negative coefficient causes a compression to be reflected as a dilatation and vice versa. Moreover, in either case the displacement in the pulse is reduced in direct proportion to the reflection coefficient.

In Fig. 57a is shown a wavelet representing the ground disturbance at the surface when a pulse reflected from an interface with an ideal reflection coefficient 1 arrives at the surface. It will be sufficiently accurate for our purpose to characterize this wavelet by a series of uniformly spaced ordinates 0, −4, −12, −8, 40, 16, −24, −8, 0 in arbitrary units. On reflection from an interface 1 having a reflection coefficient 0.5, the arriving pulse will produce half as much disturbance (assuming that the ground is a linear medium and the pulse undergoes no further change of shape in reaching the surface).

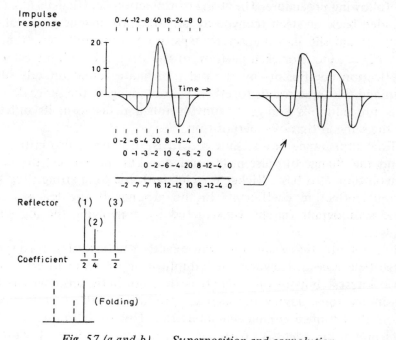

Fig. 57 (a and b).　*Superposition and convolution*

The ground disturbance due to a reflection from another interface, 2, one-half time unit 'deeper' than interface 1 (that is, $V/2$ depth units apart from 1) will start one time-unit later (since the outward going 'Ricker pulse' has to travel to and back from interface 2 in being reflected from it). Suppose the reflection coefficient of interface 2 is 0.25. Then a series of ordinates 0, −1, −3, −2, 10, 4, −6, −2, 0 displaced one time-unit in relation to the earlier series 0, −2, −6 . . . will represent the ground motion due to this reflection. Similarly a third reflector one time-unit deeper than interface 2 and having, say, a reflection coefficient 0.5, will produce a disturbance displaced two units in time. Adding (i.e. superposing) the disturbances we see that the resultant ground disturbance will be represented by the pulse in Fig. 57b with sampled ordinates 0, −2, −7, −7, 16, 12, −12, 10, 6, −12, −4, 0 (or more stringently expressed, by the ordinates series which we may use to reconstruct a first approximation of the pulse form).

Exactly the same result can be obtained if, instead of superposition, we use the following procedure. The string of reflection coefficients (1), (2) and (3) is folded back on itself (convolved, cf. the aptly named genus of plants convolvulus) and slid past in discrete steps of one time unit, across the series 0, −4, −12, . . . etc. For each position of the string we multiply each of the three reflection coefficients by the pulse ordinate found directly above it and then add the three results together. We obtain again the series 0, −2, −7, . . . This procedure is known as convolution and, as seen, its outcome is exactly the same as that of superposition.

The final appearance of a seismic record is, of course, also influenced by noise and the change of pulse in transmission but, in essence, the record is the convolution of a basic Ricker wavelet with the long string ('log') of all the various reflection coefficients in the ground. If the log is known a synthetic seismogram can be constructed by convolving the log with the basic pulse.

Ideally, a synthetic seismogram will exactly duplicate the corresponding noise-free field record. However, such duplication would be of no more than academic interest but for one fact. It is that synthetic seismograms furnish the possibility for studying the changes produced in the seismic record by changes in the assumed ground characteristics. This study in turn may enable the interpreter to identify particular geological sections. Furthermore, seismograms can be calculated for points other than the actual geophone plants. Such calculations would indicate any advantage that could accrue from additional recordings in the area and the changes one must look for in the records in that case.

The first paper on synthetic seismograms was published by Peterson *et al.* (*142*) and since then the use of the technique has been increasing. For details the interested reader may be referred to a symposium on the subject (*143*).

6.9 Continuous velocity logging (CVL)

This method of velocity determinations was first introduced by Vogel (*144*) and Summers and Broding (*145*). The idea is to measure the seismic velocity over very small intervals by lowering a suitable apparatus in a borehole. The acoustic pulse is produced by a powerful electric arc discharge in a small transmitter and the signal is picked up by a receiver only about 1−2 m away in the hole.

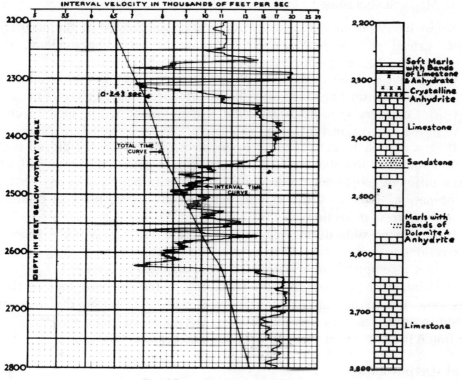

Fig. 57c.   Continuous velocity log

A part of a CV-log is shown in Fig. 57c together with the corresponding geologic section. The total travel time curve shown is obtained by integrating the interval times. The solid triangle at a depth of about 2,330 ft indicates the position of the geophone used to calibrate the CVL. The travel time to this geophone was 0.243 s.

It is evident from the CVL that the limestones in this instance give high velocities (17,000 ft/s) while crystalline anhydrite and marls with bands of dolomite, limestone and anhydrite have generally low velocities, sometimes of the order of 7,000 ft/s.

Although CVL is subject to various corrections and inaccuracies it is generally assumed that the local velocity variations evinced by it are geologically significant. The method is now part of the routine of applied seismology.

6.10 Magnetic recording

A major development is seismic instrumentation is the use of the magnetic tape instead of the photographic film as the recording medium, the galvanometers being replaced by recording heads, one for each trace.

One of the greatest advantages of the magnetic tape is that it allows a broad-band recording in which only the low frequency surface waves need be cut off. Considerably more information is therefore stored in such a record than in the conventional filtered seismograms.

If it is desired to cut off other frequencies the tape can be played back with a suitable filter and the result recorded on a paper. Moreover, outputs from different geophones can be mixed together in playback in any desired combination. This, to some extent, achieves the same object as multiple geophone arrays in Section 6.7.

It is also possible during playback to insert weathering and elevation corrections to each of the traces by separate input units and, using suitable machines, a completely automatic data processing can be arranged.

It is interesting to note that the versatility of the magnetic tape method has now led to the invention of devices which transcribe the ordinary seismogram curves onto a magnetic tape! Details of magnetic recording will be found in an excellent book by Evenden and Stone (*205*).

6.11 Deconvolution

It is evident from Fig. 57 that the effect of convolving the seismic pulse with the reflection coefficient log is to stretch or broaden the pulse. The object in seismic interpretation is the opposite, namely, to recover the original log by compressing (or deconvolving) the seismic record. The procedure is also called inverse filtering.

If s, p and l denote the seismic record, the basic pulse and the reflection log (as function of time!), convolution is symbolically expressed by the equation

$$s = p * l$$

Denote by S, P, L the respective Fourier transforms. Then (appendix, p. 201).

$$S = PL$$

where the right hand side is ordinary multiplication.

Since S, P can be expressed as polynomials in a common variable, we can obtain L (as a polynomial) by the polynomial division S/P. Computer programs exist for calculating transforms as well as for polynomial divisions (*141*). The desired log is given by the inverse Fourier transform of S/P. More generally we can write

$$L = FS \qquad (6.16)$$

where F is a 'filtering function'. A suitable form for F is, for example, $\cot(\omega t/2)\sin(m\omega t)$. If f is the inverse Fourier transform of F, the log can also be obtained by convolving f with s,

$$l = f * s \qquad (6.17)$$

In practice, deconvolution involves three principal stages:

(1) Calculation of the autocorrelation function (ACF) of a suitable section of the record, that is, estimation of the integral $(1/2T)\int_{-T}^{T} s(t)s(t+\tau)\mathrm{d}t$ for different values of τ. The procedure is computationally similar to convolution except that the sliding trace is the same as the stationary trace (nor is it folded on itself). By a 'suitable section' is meant a section that may be regarded as a random superposition of many elementary pulses each reasonably close to some average shape. T denotes the length (seconds) of the section.

(2) Choice of f. This requires, in the first place, an estimation of the desired filter F. It turns out that F is effectively the Fourier spectrum of the central portion of the ACF. Hence the need of calculating ACF in stage (1). Having estimated F we get f as its inverse Fourier transform.

(3) Application of (6.17). Since $f(t)$ is now selected and $s(t)$ is given this final step yields the desired log l. The log is obtained as a function of time but can be converted to one of depth, by appropriate assumptions about the average velocity $\bar{V}$ (migration, p. 154).

It will be seen that the entire deconvolution process is very well suited to operation on high-speed computers. It is now routinely used in the interpretation of seismic reflection data. Fig. 53d shows the section of Fig. 53c after deconvolution. The reflections are now more clearly visible due to pulse compression. It must not, however, be imagined that deconvolution always leads to a clearer display. The reason is that the inverse filter can act adversely on the noise wiggles and multiply reflected signals and may amplify them.

Considerable research has gone into designing optimum inverse filter functions and in devising rapid computer procedures for deconvolution. For details reference should be made to the special literature on the subject (*194–197*).

6.12 VIBROSEIS

This ingenious system of seismic exploration (a trademark of the Continental Oil Company) uses a vibratory instead of an impulse source. It is well known that an impulsive pulse introduces a definite band of frequencies in the earth (appendix, p. 201). In the VIBROSEIS system the signal used to introduce this band is a linear frequency sweep of the form shown in Fig. 58a. In radar technology such signals are often called 'chirps' or 'pulse compressions'. Typically, the frequency is swept from 15 to 90 c/s in about 7 seconds. The signal is introduced by truck-mounted hydraulic or electromagnetic vibrators

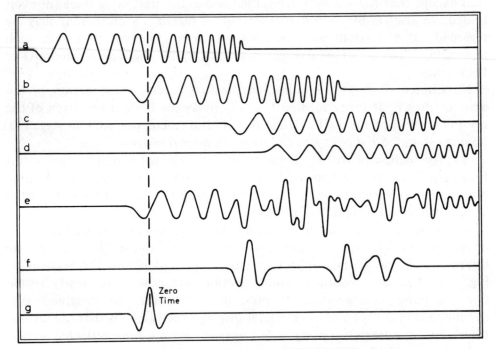

Fig. 58 (a–g).   Principle of VIBROSEIS

capable of exerting a total thrust of several tons on the earth. The input seismic pulse of the impulse sources to which the VIBROSEIS signal corresponds is given by the auto-correlation function (p. 165) of the signal. The centre of the pulse thus obtained (Fig. 58g) corresponds to the time break of the conventional methods.

The returning signal from each reflector in the earth is likewise a frequency-swept signal. Three such signals are shown in Fig. 58b, c, d. They add up to the recorded signal (Fig. 58e). To recover the reflected pulses, the total signal is cross-correlated with the input signal. Computationally this procedure is the same as convolution (p. 160) except that the cross-correlated trace is not folded but slid directly past the reference trace. For each position of it, the ordinates on the two traces are multiplied and summed. This leads to Fig. 58f which is nothing but a conventional seismic trace.

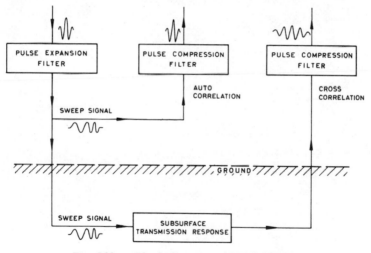

Fig. 58h.   *Block diagram of VIBROSEIS*

Theoretically, the VIBROSEIS system is the same thing as conventional seismic exploration but differs from it in the method of recovering the seismic information. However, it has several operational and practical advantages. First, the system is inherently safer and more convenient than one using explosives. Second, the sweep rates and frequencies can be optimally selected to suit the geologic conditions in the area so that power is not wasted in generating frequencies which the earth will not transmit. Finally, the energy injected into the earth can be increased almost without limit by coupling together as many vibrators as desired and making them work in phase from a master vibrator.

A complete block-diagram of the system is shown in Fig. 58h. The reader will find further details about VIBROSEIS elsewhere (*198–200*).

7·Radioactive Methods

7.1 Introduction

The geophysical methods employing radioactivity came into prominence with the demand for uranium metal in atomic reactors. However, the methods are not restricted in scope only to the search after the ores of radioactive metals or minerals associated with them (e.g. placer deposits of titanium or zirconium) but can often be used with advantage in geological and structural investigations as well.

Theoretical background.

The nucleus of an element X with an atomic number Z has a positive electric charge of Z (atomic) units and is made up of nucleons (protons and neutrons). The number of nucleons is the mass number A of the element and the nucleus is symbolically denoted as $^{A}_{Z}X$. Elements with the same Z but with different A are said to be *isotopes* of each other.

Certain nuclei disintegrate spontaneously emitting α particles (helium nuclei $^{4}_{2}He$) and β particles (electrons and positrons). This is the phenomenon of radioactivity. These emissions alter the nuclear charge, α by -2, β^{+} (positron) by -1 and β^{-} (electron) by $+1$. This means that the disintegrating nucleus is transformed into a nucleus of another element. Very often the daughter nucleus is also radioactive in its turn.

The nucleus is generally in an excited energy state after a β emission and returns to its ground state with the emission of a further particle, the γ quantum or ray. In some rare instances an α emission too is followed by a γ ray, e.g. in radium.

The γ particle is a purely electromagnetic radiation which does not alter the nuclear charge.

The disintegration of a given quantity of any radioactive element can be expressed by the formula $N = N_0 \exp(-\lambda t)$ where N_0 is the number of nuclei initially present and N the number remaining after a time t. λ is known as the decay constant and after a time $1/\lambda$ the number of nuclei present is reduced to $1/e \approx 1/3$ of the initial number.

169

A given quantity of a radioactive element is halved after a time $T = \ln 2/\lambda = 0.693/\lambda$. This time is known as the *half-life* of the element.

About 50 natural and more than 800 artificial radioactive nuclei are known. Natural radioactivity is confined principally to four *radioactive series* which start from the following isotopes of neptunium, uranium and thorium: $^{239}_{93}Np$, $^{238}_{92}U$ $^{235}_{92}U$ and $^{232}_{90}Th$. The transformations of the ^{238}U series are shown in Table 10.

TABLE 10

Radioactive disintegration of the uranium-238 series

Element	Z	Emission	Half-life	Product
Uranium-238	92	α	4.51×10^9 y	^{234}Th
Thorium-234	90	β, γ	24.1 d	^{234}Pa
Protactinium 234	91	β, γ	1.14 min	^{234}U
or		β	6.7 h	^{234}U
Uranium-234	92	α	2.52×10^5 y	^{230}Th
Thorium-230	90	α	80,000 y	^{226}Ra
Radium-226	88	α, γ	1622 y	^{222}Em (Radon)
Emanation-222	86	α	3.825 d	^{218}Po
Polonium-218	84	α, β	3.05 min	$^{214}Pb, ^{218}At$
Lead-214	82	β, γ	26.8 min	^{214}Bi
Astatine-218	85	α	2 s	^{214}Bi
Bismuth-214	83	β, α, γ	19.7 min	$^{214}Po, ^{210}Tl$
Polonium-214	84	α	1.6×10^{-4} s	^{210}Pb
Thallium-210	81	β, γ	1.3 min	^{210}Pb
Lead-210	82	β, γ	20 y	^{210}Bi
Bismuth-210	83	β, α	5.0 d	$^{210}Po, ^{206}Tl$
Polonium-210	84	α	138.4 d	^{206}Pb (stable)
Thallium-206	81	β	4.2 min	^{206}Pb (stable)

The uranium and thorium series end in the stable isotopes of lead $^{206}_{82}Pb$, $^{207}_{82}Pb$, $^{208}_{82}Pb$. The neptunium series ends in the bismuth isotope $^{209}_{83}Bi$.

Besides the members of the four radioactive series, at least 10 other naturally occurring isotopes, all of elements with atomic numbers less than that of lead, are known to be radioactive. Chief among these is the isotope $^{40}_{19}K$ of potassium with a half-life of 4.5×10^8 years. About 42 per cent of ^{40}K is transformed with an emission of a β particle into calcium 40 and about 58 per cent with the capture of an electron from the K shell of potassium into argon 40. The transformation is accompanied by high energy γ radiation of a wavelength of about 0.008 Å.

The α and β particles lose their energy in passing through matter by collisions, ionization, etc. and are brought to a virtual stop within a certain distance which is called their *range*. In air at 18°C the range of α particles is only a few centimetres, in denser substances, e.g. mica or aluminium, it is still smaller being of the order 30 μ. Even the β particles whose range is several hundred times greater are completely stopped by a thin sheet of lead or, for example, a few centimetres of sand.

The intensity of γ rays in traversing matter decreases exponentially with distance so that we cannot speak of a definite range in this case. Theoretically, γ rays could be detected across any thickness of matter but a practical limit is set by the sensitivity of the detecting instruments and the background effects due to cosmic radiation. For practical geophysical purposes γ radiation may be taken to be entirely absorbed by 1–2 m of rock.

Evidently, the α and β particles are not of much avail in geophysical field work since they will be indetectable as soon as a radioactive deposit has the thinnest cover of overburden. Thus, the search for radioactive minerals is, to a large extent, a search for places with abnormally high γ radiation. Uranium, for example, is located *indirectly* from the powerful γ radiation emitted by two products of the uranium series: $^{214}_{82}$Pb and $^{214}_{82}$Bi.

According to the quantum theory, electromagnetic radiation consists of discrete 'particles', namely photons. The energy of a photon is given by hc/λ, where h is Planck's constant, $c = 3 \times 10^8$ m/s is the velocity of light, and λ is the wavelength of the radiation. The energy is commonly expressed in electron volts (eV). One eV is the energy (1.602×10^{-19} joule) acquired by an electron in falling through a potential difference of one volt. Since $h = 6.625 \times 10^{-34}$ joule-s $= 4.14 \times 10^{-15}$ eV-s, the energy of a photon is $1.24 \times 10^{-6}/\lambda$ eV. The wavelengths of γ-rays are of the order of 10^{-11} $- 10^{-12}$ m(0.1–0.01 Å) and the corresponding energies are of the order of 0.1–1 MeV.

A radioactive series, such as that in Table 10, is said to be in equilibrium when as many atoms of any unstable element in it are being formed per second as are disintegrating. Such a series then emits a definite spectrum of γ-rays. It is customary to describe the spectrum in terms of energy levels, instead of wavelengths as for the spectrum of ordinary light. The γ-spectrum of ^{238}U contains lines at various levels from about 0.1 MeV to about

2.4 MeV. The spectrum of ^{40}K consists of a single line at 1.42 MeV. A spectral line is never perfectly sharp but has a finite, although usually small, energy width.

By measuring the energy emitted at a level it is possible to determine the amount of the particular γ emitting nuclide present in the sample, if at least this nuclide is in radioactive equilibrium with respect to its parent and daughter. If the whole sample is in radioactive equilibrium and if energy lines from other nuclides or other radioactive series are not overlapping the measured line, the content of the parent element of the series (U, Th) can be assessed by comparison with a standard of known content. Conditions are never as ideal in measurements on rock outcrops, but postulating radioactive equilibrium the measured intensities can still be converted to U or Th content, which then is referred to as equivalent content (eU, eTh).

7.2 Radioactivity of rocks

Minute traces of radioactive minerals are present in all igneous and sedimentary rocks, in oceans, rivers and springs, in oil, and in peat and humus.

TABLE 11

Radioactive contents

	U g/t	Th g/t	^{40}K %
Basalt	0.9	4.2	0.75
Diabase	0.8	2.0	
Granite	3−5	13.0	4.4
Sediments	~1.0	5−12	
Limestone	1.3	1.1	
Oil	100		
Ocean water	0.00015−0.0016	0.0005	
	%	%	
Uraninite	38−80	0−0.3	
Carnotite	50−63(†)	0−15	
Thorianite	23−26	53−57	
Monazite	0.02−0.7	3.5−16.5	

† Content of uranium oxide.

The average amounts of U, Th and ^{40}K in a few materials of the earth's crust are shown in Table 11. From this we see that there is a large difference between the radioactivity of basalts and granites. Moreover, the latter have a remarkably high content of ^{40}K. This fact is of great consequence because granites are very common rocks and the γ radiation from their potassium produces a radioactive background which may make it difficult to locate uranium and thorium ores. Sometimes the radioactivity of potassic feldspars in pegmatite dikes may be misinterpreted as being due to a concentration of uranium and thorium.

The high uranium but low thorium content of oil is also striking.

7.3 Radiation detectors

The α, β and γ radiations are detected by their ionizing action. In geophysical work only the γ rays can normally be detected because the α and β particles are easily stopped by matter.

One common type of detector is the Geiger counter. It consists of a glass tube containing some gas (usually argon with a small amount of alcohol or amyl acetate) with cylindrical cathode round a wire anode. The electrodes are kept at a high potential difference. When a γ ray passes through the gas it produces ions which are accelerated by the field and produce further ions. The momentary current passing through the tube can be amplified and registered on a meter or heard as 'click' in a pair of headphones.

The Geiger counters respond to only 1 per cent or less of the incident γ rays. On the other hand they register practically all *corpuscles* in the cosmic rays.

A more efficient type of detector is the scintillation counter. This utilizes the fact that certain crystals such as thallium-activated sodium iodide, anthracene, para-terphenyl, etc. scintillate when they absorb γ rays. The scintillations can be detected by a photocathode in a photomultiplier tube and recorded suitably after amplification.

The scintillation counters are almost 100 per cent efficient in detecting γ rays. Their sensitivity to cosmic rays is about the same as that of Geiger counters so that their relative response to γ rays is much higher.

Detectors are constructed in two basic modes. The *differential spectrometer* records only radiation falling within predetermined upper and lower energy limits. If the limits are very close together (say a few tens of keV at the most) the spectrometer is said to respond to a channel or a line, while for wider separation of limits (several hundred keV) it is said to respond to a window. The *integral spectrometer* is set to exclude radiation below a predetermined energy level, the threshold, and records all radiation having photon energies greater than this level. In some instruments the threshold can be varied. There exist also mixed-mode instruments. The choice of the spectrometer will depend to a large extent on the purpose and the requirements of a survey. Window and threshold spectrometers are more rugged, more rapid and require less checking than channel spectrometers. But the latter allow precise determinations of the various nuclides. Under controlled conditions determinations within a few g/t U or Th are possible. Channel spectrometers are, however, expensive.

Field procedure.
Radioactive surveys may be made either as 'spot examinations' or, more systematically, on a grid pattern of lines and points with the observer walking along pre-laid lines holding the detector some 40–50 cm above the ground. They can also be carried out from a car by mounting the detector on the roof provided the roads are unsurfaced, for the radioactivity of the materials in 'metalled' roads vitiates the observations.

The intensities are often recorded as counts-per-minute but most instruments are nowadays calibrated in milliröntgens-per-hour. The röntgen is the quantity of γ(or X) radiation which produces 2.083×10^9 pairs of ions per cm^3 of air at NTP. It corresponds to an energy absorption of 83.8 ergs per gramme of air (8.38×10^{-6} joules/Kg).

The intensities recorded are the integrated effects of the radiation from a finite area on the ground. In foot-surveys most of the intensity comes from within a circle of about 3 m radius while in car-borne surveys about 90 per cent of it comes (because of the greater detector height) from within a radial distance of 15–20 m.

Topographic irregularities, absorption and scattering of radiation in the earth materials, the dispersion of radioactive materials due to weathering, etc. and the 'background radiation' are some of the factors affecting instrument readings. These must be carefully considered in the interpretation.

The background is due mainly to cosmic rays, potassium 40 and the minute quantities of uranium and thorium which are almost always present in rocks. It may vary from one area to another as well as within a single area. A reading cannot be considered significant in general unless it is 3–4 times the background.

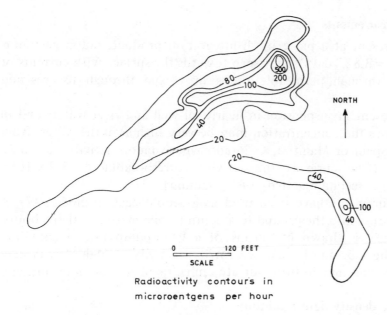

SCALE

Radioactivity contours in
microroentgens per hour

Fig. 59.   Radioactive survey (Ref. 146)

An example of a radioactive survey is shown in Fig. 59, after Moxham (*146*). The area (Pumpkin Buttes area, north-eastern Wyoming, U.S.A.) is underlain by sandstone, shale and coal of Paleocene and Eocene age. Deposits of secondary uranium are found in 'rolls' and disseminated in several sandstone layers of the Eocene.

Contrary to the above example of high radioactivity over ores, Zeschke (*201*) has reported that 20 deposits of manganiferous ore in south-eastern Europe showed a systematic *decrease* in activity as one approached the outcrops. Above the outcrops the activity was practically nil (i.e. = background). Apparently, the ore deposits have been formed by descending solutions, and the water that flowed through the openings now occupied by ore probably leached most of the radioactive material from the surrounding marble rocks which, it is presumed, had originally a higher radioactivity than at present. This example indicates that the interpretation of radioactivity surveys, like that of all other geophysical surveys, must be done in conjunction with geological information.

7.4 Radon measurements

If radium is present at a place its disintegration product, radon gas (an α emitter with $T = 3.825$ days), may seep towards the surface with currents of water, diffuse through permeable rocks or escape through fissures and cracks.

Radon is present in suspension in nearly all well and river waters and in oil. In some cases the concentration may be very high as in the waters from Valdemorillo, Spain or Manitou, Colorado whose radon activities are 0.22 and 0.03 $\mu c/l$. (One curie is the activity corresponding to 3.7×10^{10} disintegrations per second = activity of 1 g radium).

Radon measurements have been used as a geophysical method (*147*). A small tube is thrust into the ground to a depth of one metre, or thereabouts. A sample of gas is drawn by means of a hand-pump into a container connected to the tube and is analysed for its α activity. A high activity may be due to radon and may in turn indicate faults, fissures, uranium veins, etc.

7.5 Radioactive density determinations

The absorption of γ-rays has been employed in recent years for determining soil density in foundation investigations and hydrological problems. For this purpose an artificial radioactive isotope such as ^{60}Co ($T = 5.26$ years) or ^{137}Cs ($T = 33$ years) is used. The former emits γ rays during beta disintegration, the latter changes to an excited ^{137}Ba which in turn emits γ rays in returning to the ground state.

A metal probe with the γ source at the end is driven vertically to a depth (z) of about 50–100 cm. The count of a detector placed on the surface at various horizontal distances (x) along a line is noted. The soil will absorb γ rays so that the count will depend upon its density ρ.

If the detector is assumed to be very small it can be shown that

$$N = \frac{N_0}{4\pi r^2} \mu \exp (-\mu r) \qquad (7.1)$$

where N_0 is the total number of γ quanta emitted, N the number reaching the detector, $r = \sqrt{(x^2 + z^2)}$ the source-receiver distance and μ an absorption coefficient.

The plot of $\ln(r^2 N)$ against r will evidently be a straight line with a slope μ. It can be proved however that $\mu = \mu' \rho$ where μ' depends only upon the energy of the radiation. The mean values of $1/\mu'$ are 18.4 g/cm^2 for ^{60}Co and 14.2 g/cm^2 for ^{137}Cs but are affected slightly by test conditions. For the examples shown in Fig. 60 (after Homilius and Lorch) using ^{137}Cs $1/\mu'_s = 15.33$ g/cm^2 (clay) and 15.65 g/cm^2 (sand).

Density values accurate to 1 per cent or better can be obtained with this method whose chief advantage is that the density can be determined *in situ*, without having to take samples.

A rigorous and detailed discussion of the method has been given by Homilius and Lorch in two interesting papers (*148, 149*).

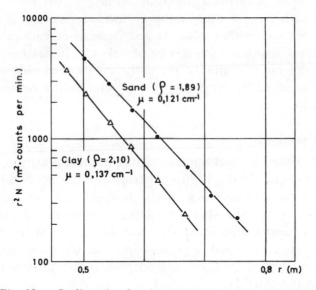

Fig. 60.   Radioactive density determinations (Ref. 148)

8·Airborne Geophysical Methods

8.1 Introduction

The magnetic, electromagnetic and radioactive methods have been adapted to geophysical measurements from the air. Airborne work has certain advantages. Firstly, on account of the high speed of operations an aerial survey is many times cheaper than an equivalent ground survey provided the area surveyed is sufficiently large and secondly, measurements can be made over mountains, jungles, swamps, lakes, glaciers and other terrains which may be inaccessible or difficult for ground surveying parties.

Compared with ground work, airborne measurements imply a decrease in resolution which means that adjacent geophysical indications tend to merge into one another giving the impression of only one indication. Besides, there is often considerable uncertainty about the position of airborne indications so that they must be confirmed on the ground before undertaking further work like drilling.

8.2 Magnetic methods

Conventional ground magnetometers require a stable platform and cannot be used in an aeroplane. In the 1920's experiments were made with an airborne earth inductor but the accuracy of the method was not sufficiently high. During the war years the flux-gate detector (Chapter 2, Section 4) was developed as an anti-submarine device and this is one of the most common types of magnetometer used at present in aerial work. Another instrument that has been adapted to the same purpose is the proton free-precession magnetometer (Chapter 2).

The sensitive element of the magnetometer is towed behind the aeroplane at the end of a cable (in order to minimize the effect of the plane's magnetism) and its output is fed into the recording equipment which is in the plane. The readings along the flight line are registered continuously on a running paper strip.

In the flux-gate magnetometer the sensitive element (high permeability plates) are oriented automatically in the direction of the earth's field so that the instrument records the *total* intensity. The use of a gyroscope would allow the measurement of the vertical or the horizontal component but the accuracy is then not better than about 50–100 gammas compared with about 1 gamma for the total intensity.

The proton magnetometer necessarily measures the ambient field, that is, in the present case, the earth's total field.

The interpretation of airborne magnetic anomalies follows more or less the same pattern as in ground work but is mainly on qualitative lines. Isoanomalous charts can be prepared as in ground work if the measurements are made along sufficiently closed spaced flight lines. In quantitative calculations the chief difference arises due to the fact that formulas such as (2.6) and (2.7) in Chapter 2 must be replaced by the corresponding formulas for the total intensity. Such modifications have been considered by Henderson and Zietz (*150*) and by de Géry and Naudy (*151*).

8.3 Electromagnetic methods (continuous wave)

These airborne electromagnetic methods are modifications of the moving source-receiver system described in Chapter 5, Section 5. Three principal types have been proposed: (1) Helicopter and wing-tip systems, (2) dual frequency system and (3) two-plane rotating field system.

(1) *Helicopter.*

In its typical version this system employs vertical, coaxial transmitter and receiver coils with the dipole axis parallel to the line of flight (cf. Fig. 44b). The coils are often compact ferrite-core coils with a diameter of approximately 10 cm and are mounted about 10–20 m apart at the ends of a very rigid boom installed underneath a helicopter. The boom is carefully designed to minimize the motion of the coils relative to each other so that spurious signals in the in-phase component are eliminated.

The system is flown at a height of about 45–50 m and the in-phase and out-of-phase field components at the receiver are continuously registered (usually in parts per million of the primary field at the receiver) on the automatic recorder. Typical anomalies from subsurface conductors range from a few hundred to a thousand parts per million, and conductivities are estimated semi-quantitatively by the 'In-phase/Out-of-phase' response ratio. The shapes of the anomalies can be visualized by the same type of arguments as for ground systems (p. 116 or p. 133).

The coils can also be mounted on the wing-tips of a small aircraft. In that case they are situated in the same vertical plane (instead of being coaxial as in the helicopter system) with their axes in the flight direction.

The practical depth penetration of the two systems is of the order of 20–30 m below ground level (about 75 m below flight level). Some recent experiments in Sweden show however that, with optimum spacing, it may be possible to obtain a much greater depth penetration.

(2) *Dual frequency.*

The fixed-wing aircraft has certain advantages (longer range, higher survey speed, greater pay-load, etc.) over the helicopter. However, the desired constancy of the coil distance (less than about 1 cm in about 15–20 m) is difficult to achieve with the coils on the wing-tips.

Now, the out-of-phase component of an electromagnetic field is a purely secondary phenomenon and is independent of the variations in the coil distance.* In the dual frequency system only the out-of-phase component is recorded and this at two frequencies, one low (400 c/s) and one high (2,300 c/s). The response ratio 'Low/High' instead of the ratio 'In-phase/Out-of-phase' provides a measure of the conductivity of the anomalous body.

The transmitter is at the aeroplane while the receiving coil is towed behind in a 'bird' at the end of a cable, some 150 m long. The plane must be flown at a height of at least 120–150 m in order that the bird may not hit the tree tops and get lost. Over undulating terrain and in 'bumpy' weather it might be necessary to fly it still higher.

Measurements of the in-phase component are out of the question due to the violent movements of the receiver bird.

The response of the dual frequency system to superficial conductivity variations is of the same order as its response to typical ore conductors, which of course, are the principal objects of electromagnetic prospecting. This introduces a 'noise' in the record in addition to the instrumental noise

* This is strictly true of the phase-shift and only approximately (but to a very high degree) of the out-of-phase component.

and leads to very poor depth penetration – probably only about ten metres below the ground surface – unless the system is flown very low which however is rarely possible for technical reasons.

Furthermore, the out-of-phase response of very good conductors, e.g. many sulphide ores, is extremely small at 400 as well as 2,300 c/s and falls within the total noise level of the system. Clearly, such conductors cannot be detected by out-of-phase measurements alone. This disadvantage may, however, be offset in areas with conducting host rock and/or overburden because good conductors may collect the phase-displaced currents induced in the rock or the overburden and give rise to a measurable electromagnetic field. (This is often the reason behind the strong imaginary component anomalies obtained over grounded rails, metal pipes etc, which, in themselves, represent almost infinitely good conductors to electromagnetic prospecting systems. Their intrinsic imaginary component responses are zero.)

The shortcomings of the dual frequency method have been discussed in detail elsewhere (*152*).

(3) *Rotating field.*
This method was devised in Sweden with a view to overcoming the disadvantages of the systems under (1) and (2), namely the small depth penetration and, in (2) also the loss of information due to the high noise level and due to the inherent impossibility of in-phase measurements.

The transmitter of the system consists of two coils, one horizontal and the other vertical, fed by alternating currents of the same amplitude and frequency but with a phase difference of 90°. The primary electromagnetic field at any point in the surrounding space is therefore a rotating elliptically polarized field.

The receiver consists likewise of two mutually perpendicular coils and is placed at a distance, say a, from the transmitter (Fig. 61a). Along the line joining the transmitter and receiver in Fig. 61a the ellipse of polarization is a circle.

It is evident that if m is the dipole moment of either transmitting coil the primary field acting on the horizontal as well as the vertical receiving coil has an amplitude $m/2a^3$. The voltages induced in the two receiver coils can be balanced against each other after shifting the phase of one of them, say that of the vertical coil voltage, by 90° so that the reading of a meter or a recorder is normally zero.

When a secondary field from a subsurface conductor acts on the receiver there is an unbalance since the field will in general affect the two receiver coils unequally. The net voltage in the vertical coil is added (after shifting its phase by 90°) to the voltage in the horizontal coil. The resultant has two components, one in phase with the primary field in the horizontal receiving coil and the other out of phase with it. These unbalanced components can be measured in terms of volts, amperes, gauss or any other convenient unit by suitably calibrating the deflection of the recorder connected to the receiver.

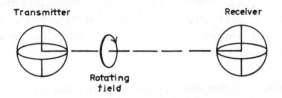

Fig. 61a.   *Principle of the airborne rotating field system*

It is, however, preferred to express them as percentages of the primary voltage induced in either receiver coil when a has a standard, specified value a_0.

It will be noticed that in contrast to the other airborne systems the rotating field system does not require a direct cable connection between the transmitter and the receiver. Also, the balance of the primary fields is independent of the variations in a since these affect the fields on both receiver coils in exactly the same way.

It is therefore possible to place the transmitter in one aeroplane while the receiver is towed on a short cable (about 15–30 m) from another plane, flying in tandem at a distance of about 150–300 m. Such a system can be flown fairly low (about 60–80 m) and is not so dependent on weather conditions as the dual frequency system with its long towing cable.

Theoretically, the low flight height and the large transmitter-receiver distance should combine to give the system a depth range of more than 100 m below ground (*153, 154*). However, the response of superficial horizontal conductors (e.g. soil layers) to the rotating field system is often very complicated, masking the anomalies of deeper conductors and making the interpretation difficult. According to tests in areas of high surface conductivity (East Africa) the practical penetration of the rotating field system in these areas appears to be only slightly better (40—50 m) than that of the wing-tip systems (*203*) although in areas of low surface conductivity (e.g. some precambrian shield areas) the predicted figure may perhaps be approached.

8.4 Airborne AFMAG, Radiophase, Turair and INPUT

AFMAG.
This system (section 5.11) has also been adapted to airborne work (*118*). A receiver consisting of two mutually perpendicular search coils, each making an angle of 45° with the horizontal, is towed behind an aircraft with the axes of the coils and the line of flight in the same vertical plane.

The flight direction is chosen perpendicular to the geologic strike in the region. Since the natural magnetic fields tend to be polarized perpendicular to the geologic strike the mean polarization vector lies then in the vertical plane through the flight line and the coil axes. If the vector has a tilt α from the horizontal, the voltages induced in the coils are proportional to $\cos(45° - \alpha)$ and $\cos(45° + \alpha)$. The difference of the two voltages is recorded in such a way that the deflection of a pen is approximately proportional to α. As in the ground AFMAG the tilt is recorded at two frequences (150 and 510 c/s) so that relative estimates of the conductivities of anomalous conductors can be made.

In a way the AFMAG may be considered to be an ordinary electromagnetic method with the source (thunderstorms) removed to infinity. A large depth penetration is therefore to be expected for the method and, in fact, trial flights are claimed to have shown a possibility of a penetration of several hundred metres below ground.

Radiophase (VLF).
This is a system for combined H and E mode measurements from the air on the VLF band (section 5.12). A block diagram of the system is shown in Fig. 61b.

There are two electrical antennas, one vertical ('whip') and the other

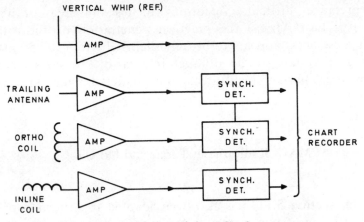

Fig. 61b.   Block diagram of the radiophase system

horizontal ('trail antenna'). Similarly there are two mutually orthogonal coils for picking up the magnetic field.

Since the amplitude of the vertical component (E_z) of the electric vector **E** is 100–200 times larger than that of the horizontal component (E_x), small departures from orthogonality between the two antennas cause large variations in the amplitude of the total signal picked up by the trail antenna. Hence a direct measurement of the ratio between the whip and trail antenna voltages is out of the question. What is measured instead is the amplitude of the component of the trail-antenna voltage that is 90° out-of-phase with the whip-antenna voltage, in terms of the latter's amplitude. It is easily seen that the measured quantity is $(E_x \cos \gamma / E_z) \sin \phi$ where ϕ is the phase difference between E_x and E_z, and γ is the instantaneous angle made by the trail antenna with the horizontal. It is not possible to relate this quantity to the tilt of the polarization ellipse without a knowledge of γ, and since this is impossible in airborne work, we cannot resort to (5.8) for defining ρ_a. The method is therefore only of semi-quantitative use for mapping conductivity variations.

The *H*-mode measurements are carried out in a similar manner. The components of the signals in the two coils that are 90° out of phase with the whip-antenna voltage are measured and r.m.s. summed to give a measure of the amplitude of the secondary magnetic field. The measure is not theoretically exact.

As in ground work, the scope of the airborne VLF-method is primarily limited, on account of the high frequency involved, to the mapping of superficial conductivity variations e.g. of soils, moraine cover, permafrost areas etc.

Turair.

This system is based on the principle of the Turam ground method (p. 113). It is a semi-airborne system in that the transmitter (large loop or long cable) is laid on the ground and traverses are made across it with the two Turam receiver coils towed from an aircraft. The corrections for the normal ratios are made by graphically removing the 'regional' gradient that appears on the profile record. They could, of course, be made by computation if the aircraft position with respect to the transmitter is known but this would probably add greatly to the cost of the method. The interpretation of the anomalies obtained is exactly as in ground work, a remark that applies to all airborne work.

INPUT.

Transient pulses (section 5.9) have been employed for aerial electromagnetic work. In the *IN*duced *PU*lse *T*ransient system a wire is strung from the aircraft nose, around each wing-tip and beneath the tail, thus forming a large loop transmitter. Pulses of some 2kW are sent through this horizontal loop at a repetition rate of several hundred times a second. A small 'bird' towed on a cable about 150 m long houses the coil which receives the secondary decaying signal from the ground. The signal is sampled (between the transmission pulses) at several preselected decay times (channels) from about 100 μs after current cut-off to about 2000 μs. The results can be expressed as ratios of the signals on different channels to, say, the signal in channel 1. A rapid decay (low late-channel signal) indicates a poor conductor in the ground, a slow decay a good conductor.

An advantage of the INPUT system is that since measurements are made during the time the primary field is cut off, the transmitter-receiver orientation geometry is of no consequence. Hence the sensitivity of the (broadband) receiver can be increased very considerably so that extremely weak signals can be detected. However, at the same time, we have no absolute reference for the strength of the recorded signals. In this the INPUT system resembles purely phase-measuring systems (p. 180). It is, in principle, a multi-frequency phase-measuring system, the later sampling times corresponding to successively lower frequencies. The lowest frequency that is effectively exploited cannot, of course, be less than the repetition rate of the pulse and the highest cannot be greater than the reciprocal of the time (after power cut-off) at which channel 1 is sampled.

8.5 Radioactivity

As with the other methods, airborne radioactivity work differs from the ground work mainly in respect of the operational procedure. The general principles of measurement and interpretation are the same in either case.

The intensity of the radiation from radioactive sources such as uranium veins decreases rapidly with the height of the observation point above the earth's surface. The effect is due, in part, to geometrical divergence and, in part, to absorption and scattering in the air. Needless to say, scintillometers of the highest sensitivity are needed in airborne work and flight heights must be kept relatively low (50–70 m). The United States Geological Survey, for example, has used instruments with up to six thallium-activated NaI crystals (some ten centimetres in diameter and five centimetres thick) coupled in parallel. Such a detection system is reported to be sensitive to differences as small as 10 parts per million in the uranium content of rocks.

The detectors are generally shielded (except for the face of the crystal which is directed downward) from cosmic rays by means of a lead shield. The radiation intensities are recorded continuously and the deflections on the recording chart are calibrated by flying the detector over naturally occurring or artificially placed radioactive sources of known strength.

Considerable attention has been paid to devising methods for quantitative estimations of the radioactive content of rocks from the strength and nature of airborne anomalies and great progress has been made in this direction. Moxham (*146*) states that the content of sources with large dimensions, e.g. marine phosphate deposits, can be determined within a few thousandths of a per cent uranium (or its equivalent). The accuracy decreases with the areal extent of the source.

8.6 Position location

The interpretation of airborne geophysical anomalies and the subsequent ground follow-up require a knowledge of the altitude of the aircraft and its position in the horizontal plane.

The altitude is measured by a barometric or a radio-altimeter and is recorded continuously alongside the traces representing the results of the geophysical measurements. The position of the aircraft in the horizontal plane is much more difficult to determine. If detailed large-scale maps or good aerial photographs are available the following method is commonly used for this purpose.

The line to be flown is drawn on the map or the photograph and the pilot takes his bearing on a suitable distant landmark on the line. A navigator sitting beside the pilot watches for details on the flight lines such as rivers, lakes, brooks, railways, houses, etc. At intervals, when flying over a clearly identifiable feature he makes a mark on the map and at the same time presses a signal button which puts another mark on a trace on the recording paper. The succession of marks on the map and the recording paper can be correlated at the end of the flight. The navigator also informs the pilot of any adjustments necessary in the course of the plane.

If maps or air photographs are not available a camera is installed in the plane and exposures are made at regular intervals during flight. The number and the instant of an exposure are, of course automatically registered on the recording paper. The interval between exposures is usually such that a certain amount of overlap between two adjacent photographs is obtained.

The above position location systems function satisfactorily only so long as there are a sufficient number of distinct topographic features in the area. In flying over dense jungles, deserts, seas, etc. the paucity or absence of such features severely limits the value of photographic methods and some form of radio location must be used. Some of these methods employ radar pulses while others are operated on continuous radio frequency waves. The general principle of the latter methods (Decca, Raydist, Lorac, etc.) is to send out radio waves simultaneously from two transmitter stations at fixed and precisely known locations and measure the phase difference between them at the receiver station (aeroplane). The phase difference is essentially a measure of the difference between the travel times of the two electromagnetic waves and is therefore proportional to the difference of the transmitter distances from the aeroplane.

Recently another method of radio location using the Doppler effect has been introduced and seems to offer some advantages.

Four radio beams directed towards the ground are sent out making a small angle with the vertical from a transmitter in the aeroplane, two in the forward and aft directions and two sideways. The Doppler frequency shifts of the reflected waves due to the motion of the aeroplane are made to yield the deviations of the plane from the flight line. Thus, starting from a given station and a given initial bearing the actual flight path can be reconstructed from the continuous record of the Doppler shifts.

9 · Miscellaneous Methods and Topics

As the requirements of modern oil and mining exploration, hydrological investigations and civil engineering have grown, a number of special techniques suited to particular problems have been proposed and applied with varying degrees of success. The purpose of this chapter is to mention a few of these techniques and also discuss some topics which are of common interest in all geophysical surveys.

9.1 Borehole magnetometer

The measurement of electric resistivity, self-potential and elastic wave velocities in boreholes has already been mentioned (pp. 97, 140, 162). It may be added that magnetic intensities too can be measured in a borehole by lowering a flux-gate or a proton magnetometer in it.

The borehole magnetometer has been used chiefly in iron-ore prospecting as an auxiliary instrument. If, for instance, a borehole has failed to encounter an expected ore zone, magnetic measurements can frequently reveal whether the zone is present in the vicinity of the hole and also indicate its distance. Weak and erratic mineralization often makes it difficult or impossible to establish the precise limits of a magnetic impregnation zone by an inspection of the drill cores alone. In such cases, the borehole magnetometer may be of considerable assistance.

The instrument can also be employed sometimes for correlating zones of the same grade in different parts of a magnetite ore deposit.

9.2 Gamma-ray logging

This technique utilizes the natural radioactivity of rocks and is used chiefly for correlating sedimentary strata in petroleum prospecting. The apparatus consists, in principle, of a gamma-ray detector and its preamplifier suspended in a borehole by means of a waterproof electrical cable. The output of the

189

detector is further amplified on the surface and recorded continuously as the detector is lowered (or raised) in the hole.

Generally speaking, shales and shaly sandstones show a very high radioactivity (the highest in sedimentary rocks) while salt, coal, anhydrite, limestones, quartz sands, etc. are weakly radioactive. Thus in a gamma-ray log the peaks in the intensity will correspond, in general, to shales while the lows will indicate the presence of limestones, salt, etc.

The gamma-ray log has also been used in uranium prospecting (*155*).

9.3 Neutron logging

There are two borehole methods known as neutron–gamma and neutron–neutron logging which employ neutrons (*156, 157*). A convenient source of neutrons often used is a mixture of radium and powdered beryllium. The beryllium is bombarded by α particles from the radium and fast neutrons are produced according to the reaction

$$ {}^9_4\text{Be} + {}^4_2\text{He} \rightarrow {}^{12}_6\text{C} + {}^1_0\text{n} + \text{energy} $$

In collisions with nuclei neutrons are gradually slowed down until they reach thermal velocities. Hydrogen nuclei are particularly efficient in producing such 'thermal' neutrons and then capturing them according to the reaction

$$ {}^1_0\text{n} + {}^1_1\text{H} \rightarrow {}^2_1\text{H} + \gamma $$

with the production of γ radiation.

In rocks, hydrogen nuclei are present in oil, water, natural gas, etc. and the γ radiation to which they give rise can be detected by lowering a neutron source just ahead of a gamma-ray counter.

In the neutron–neutron method the intensity of the neutrons scattered by the hydrogen nuclei, rather than the intensity of the gamma radiation due to their capture, is detected. Neutrons do not produce appreciable ionization so that a special device is needed for their detection. One such device is a Geiger tube filled with boron trifluoride gas. The neutrons react with the boron transforming it into lithium and releasing an α particle which in turn ionizes the gas and reveals the presence of neutrons.

In passing through matter with a high hydrogen content the neutrons are slowed down and captured by the hydrogen nuclei at a very small distance from the source. On the other hand, if the hydrogen content is low they travel a relatively large distance before reaching thermal velocities.

The number of neutrons arriving at the detector will be less when hydrogen is present than when it is absent. A decrease in the neutron response therefore indicates the presence of hydrogen in the rock and it may be assumed that this in turn is due to oil, water or gas, all of which have an abundance of hydrogen nuclei.

A neutron log and, for comparison, the resistivity log in the same sedimentary column are shown in Fig. 62.

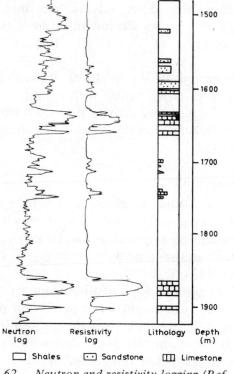

Fig. 62.   Neutron and resistivity logging (Ref. 56)

9.4 Geothermal methods

Owing to the radioactivity of the crustal rocks heat is being continuously transported to the surface of the earth from its interior at a mean rate of about 1.2×10^{-6} cal/cm^{2}s (50 mW/m^2). The lateral variations of this heat flow over the surface of the earth are small. If there are any appreciable local variations they are superimposed on the thermal effects due to vegetation, microclimates, etc. The latter are so large that surface temperature measurements cannot (as a rule) be used for deducing the thermal conductivity of rocks at depth or for determining the position of structures such as buried domes, anticlines, etc. They have, however, been successfully applied in finding fissures and cracks along which convective transfer of heat has taken place from the depth through the agency of water (*158*).

Also, Poley and van Steveninck have recently shown in an excellent paper (*202*) that temperatures only about 2 m below the ground surface are substantially free from the above effects and measurements in short drillholes are capable of revealing shallow salt domes and similar structures by anomalies of the order of 1−2°C (50−100 times the measurement accuracy!).

Contrary to the behaviour of the total heat flow, the vertical gradient of the temperature in the earth varies within wide limits (5−70°C/km) depending upon the thermal conductivity of rock formations and temperature logs in deep boreholes can be used with advantage in correlating stratigraphic horizons (*62*).

9.5 Geochemical prospecting

Indications of oil, ore and other minerals can be sought by chemical analyses of the solid, liquid, and gaseous substances of which the earth's crust is composed. Such geochemical prospecting can be carried out in the bedrock, in loose overburden or in the uppermost layer of the earth's crust, namely the surface soils (*159, 162*).

The analyses of the gases slowly diffusing from very great depths to the surface of the earth has been employed in petroleum prospecting for quite some time. In this case the contents of hydrocarbons like methane, ethane, propane, etc. (which are intimately associated with oil) are determined in samples of gas collected just below the surface of the earth.

In mineral prospecting the object is to determine the metallic contents at different places on the earth. Metal ions can migrate through a number of agencies like weathering, leaching, wind transport, etc. several tens or even hundreds of metres from the parent ore body and create 'haloes' of mineralization around it. The metal concentrations in such haloes are in general very small and refined colorimetric and spectrographic methods are needed for reliable determinations.

The tracing of haloes is reported to have led to the discovery of a large number of mineralized zones in different parts of the world.

The phenomenon of the migration of metals within the earth's crust is an exceedingly complex one. Nevertheless, great progress has been made in its study, particularly in the U.S.S.R. where the subject has reached its highest development at present (*159*). The importance of geochemical prospecting will be apparent from the fact that many metals like molybdenum, tungsten, vanadium, etc. which are of vital importance in modern industry, occur in the earth's crust in quantities far too small to appreciably alter the physical properties of the host rock so that their detection by geophysical means is out of the question.

9.6 Optimum point and line spacing

It is clear that the amount of information that can be extracted about subsurface features from geophysical measurements will be greater the denser the network of the observation points. A maximum amount of information will be available when the points are infinitely close to each other. However, this ideal cannot be realized, firstly due to the finite size of any instrument and secondly due to economic and practical considerations. Therefore a balance must be struck between the extent of information desired and the amount of detail to be mapped.

Geophysical observations are most conveniently made at a number of consecutive points along a set of parallel straight lines. It is advantageous to stake the lines perpendicular to the known or presumed geological strike, which is the direction of the trace made on the earth's surface by inclined strata.

The geological conditions tend to be uniform over relatively large distances along a line in the strike direction but may vary considerably within short distances in the perpendicular direction (Fig. 63). If the lines of measurement are normal to the strike their mutual distance may be kept large while the geophysical observations are taken along each of them at relatively close intervals.

The distance between observation points must be adjusted to the anticipated depth of the feature one is seeking. The anomalies of shallow features are narrow while those of deep-lying features extend over greater distances. As a rule geophysical observations cannot be expected to yield

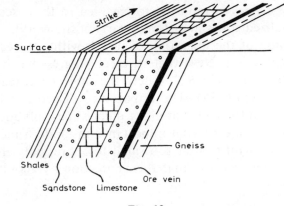

Fig. 63

information about features whose depth is much smaller than the distance between the observation points.

The optimum distance between the lines is dependent upon the length of the subsurface feature, or more correctly, upon the estimated length of its measurable surface anomaly. If the distance is too large there may be no line which crosses a short anomaly and the corresponding feature may be undetected. On the other hand, if the line distance is too small there may be quite a number of lines crossing a long anomaly. In general, such lines tend to add to the cost of the survey without providing very much additional information.

Mathematically, the problem of optimum line spacing is that of finding the probability of detecting a geophysical anomaly of given dimensions by surveys with a particular line spacing. This is actually an old problem in 'continuous geometric probabilities' (*160*) whose solution is well known. For example, if the length of an unknown anomaly is L, the probability that at least one of a set of lines with a spacing $S(\geqslant L)$ crosses the anomaly is $2L/\pi S$. If a number of ore bodies each of which produces an anomaly of length L are randomly distributed in an area the probability of detecting them by a survey with a line spacing $S = L$ is 0.636.

For $S \leqslant L$ the probability is

$$\frac{2L}{\pi S} [1 - \sqrt{(1 - S^2/L^2)}] + \frac{2}{\pi} \cos^{-1} \frac{S}{L}$$

If the length of the ore bodies is, say, 100 m the probability of detecting them all with a line spacing of 20 m is 0.974.

Other aspects of optimum line spacing have been considered by Agocs (*161*).

9.7 Composite surveys

The choice of the method for a geophysical survey is guided by a number of considerations such as the object of the survey, the geology and topography of the area to be investigated and the type of information sought about the subsurface. The last mentioned factor is, of course, of fundamental importance.

It is frequently advantageous to use two or more suitable methods within a given area. Such a composite survey gives additional information about the physical properties of the subsurface and helps (in combination with the geological knowledge) to reduce the uncertainty inherent in the interpretation of geophysical data.

For example, an electromagnetic anomaly may indicate an electric conductor but sometimes the conductor may be so good that the strong eddy currents in it will screen other less good conductors in the vicinity from the primary field and these will not be detected. In such a case an earth resistivity (or a self-potential) survey will often reveal their presence. A magnetic survey will further eliminate (or confirm) the possibility whether these conductors contain magnetite (pyrrhotite) or are composed purely of non-magnetic conducting minerals. A gravity survey will often help to distinguish between compact, massive ore bodies and zones of poor, disseminated mineralization since the density contrast with the host rock will be large in the first case and small in the second.

Faults, domes, anticlines and other geological structures indicated by a seismic survey can also be 'screened' by gravity observations to obtain information about the density contrasts involved, while magnetic measurements can frequently reveal diabase and gabbro dikes which may not be indicated by the seismic survey owing to their insufficient velocity contrast with the country rock.

Examples of the advantages of composite surveys can be easily multiplied. The results along one of the profiles in a composite electromagnetic, magnetic, self-potential, earth resistivity and induced polarization survey in a

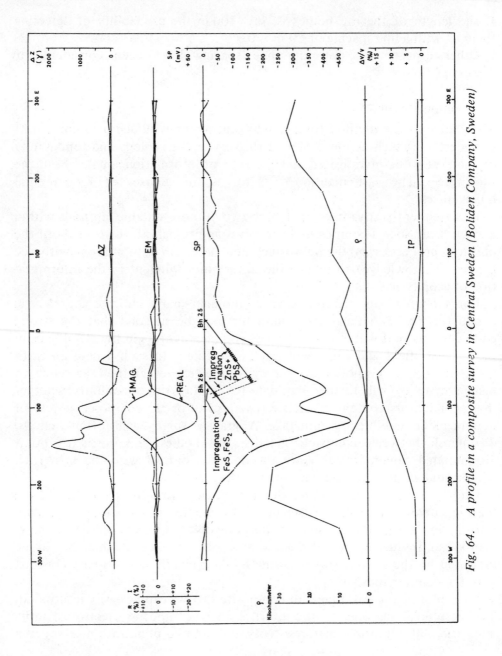

Fig. 64. A profile in a composite survey in Central Sweden (Boliden Company, Sweden)

196

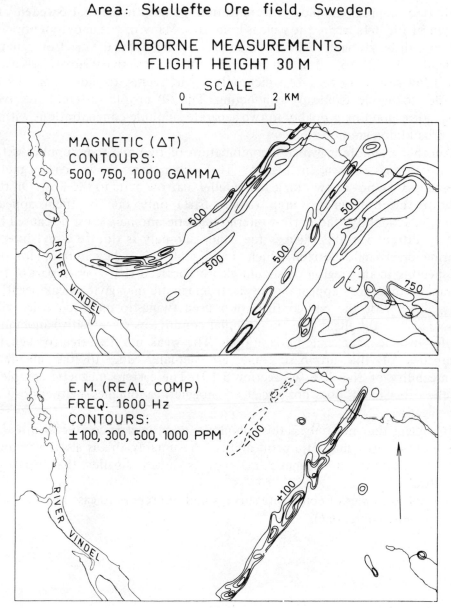

Area: Skellefte Ore field, Sweden

AIRBORNE MEASUREMENTS
FLIGHT HEIGHT 30 M
SCALE
0   2 KM

MAGNETIC (ΔT)
CONTOURS:
500, 750, 1000 GAMMA

RIVER VINDEL

E.M. (REAL COMP)
FREQ. 1600 Hz
CONTOURS:
±100, 300, 500, 1000 PPM

RIVER VINDEL

Fig. 65.   Combination of electromagnetic and magnetic surveys (Boliden Company, Sweden)

197

lead—zinc—copper sulphide mineralization district in central Sweden are shown in Fig. 64, where the scale is in metres. Many of the above-mentioned effects will be recognized in this example. For instance, the electromagnetic indication is due to a thin sulphide vein but the resistivity survey reveals a broad low-resistivity zone of which only a part seems, to judge from the IP profile, to be electronically conducting. The SP profile differentiates two tops, corresponding probably to two separate sulphide concentrations within the broad impregnation zone.

Another example in which a combination of two methods supplies extra information is shown in Fig. 65. The upper part shows aeromagnetic anomalies. We notice two long and parallel, narrow belts to the west. On the airborne electromagnetic map (lower part) only one of them appears strongly. Evidently, the north-western magnetic anomaly must be caused by a long ultrabasic dike whereas the other anomaly is due to a pyrrhotite-bearing black-shale horizon which, incidentally, is a key horizon for ore prospecting in this region. It should also be noticed that a small part of the ultrabasic dike does appear on the electromagnetic map giving weak negative anomalies. With the coil-configuration used (wing-tip mounted coils with dipole axes in the flight direction) tabular conductors give positive anomalies as, for instance, the shale horizon does. The weak negative electromagnetic anomalies of the ultrabasic dike are actually due to the magnetic permeability of the dike (cf. section 5.10). The eastern magnetic complex, on the other hand, does not give electromagnetic anomalies, neither positive nor negative.

It is clear that not only is this complex a poor electric conductor but also that it has a low magnetic permeability. Presumably this is also basic rock but its content of magnetic material is much smaller than that of the ultrabasic dike.

Several examples of composite surveys and references to case histories will be found elsewhere. (*61*).

Appendix
Fourier Transforms and
Convolution

Fourier transforms (section 5.9).

Any arbitrary function $f(t)$ (satisfying certain conditions) can be synthesized from a number of sine and cosine waves of different frequencies each with a characteristic amplitude and phase. If $f(t)$ is periodic with the period T the frequencies are discrete multiples of the fundamental frequency $1/T$. If $f(t)$ is non-periodic the frequencies are infinitesimally close to each other and range from 0 to ∞. For mathematical convenience, however, they are usually taken to range continuously from $-\infty$ to $+\infty$. The function $F(\nu)$ giving the amplitude of the wave of frequency ν is known as the amplitude spectrum. It can be obtained, as will appear below, from $f(t)$ while, conversely, if $F(\nu)$ is given $f(t)$ can be synthesized by essentially the same process.

We shall start with a periodic function $f(t)$ and express it by means of a Fourier series:

$$f(t) = \sum_{m=0}^{\infty} \left\{ a_m\, e^{i(2\pi m/T)t} + b_m\, e^{-i(2\pi m/T)t} \right\}$$

It is easily shown by multiplying both sides by $\exp(i(2\pi n/T)t)$ and $\exp(-i(2\pi n/T)t)$ and integrating from $-T/2$ to $+T/2$ that

$$a_m = (1/T) \int_{-T/2}^{T/2} f(t')\, e^{-i(2\pi m/T)t'}\, dt'$$

$$b_m = (1/T) \int_{-T/2}^{T/2} f(t')\, e^{i(2\pi m/T)t'}\, dt'$$

Hence

$$f(t) = (1/T) \sum_{m=0}^{\infty} \int_{-T/2}^{T/2} f(t') \left\{ e^{i(2\pi m/T)(t - t')} + e^{-i(2\pi m/T)(t - t')} \right\} dt'$$

Since the bracketed expression on the right hand side is an even function of *m* it is obvious that we may also write

$$f(t) = (1/2T) \sum_{m=-\infty}^{\infty} \int_{-T/2}^{T/2} f(t') \left\{ e^{i(2\pi m/T)(t - t')} + e^{-i(2\pi m/T)(t - t')} \right\} dt'$$

Moreover, since $\exp(x) - \exp(-x)$ is an odd function of *x* we obviously have

$$0 = (1/2T) \sum_{m=-\infty}^{\infty} \int_{-T/2}^{T/2} f(t') \left\{ e^{i(2\pi m/T)(t - t')} - e^{-i(2\pi m/T)(t - t')} \right\} dt'$$

Adding these two equations we get

$$f(t) = (1/T) \sum_{m=-\infty}^{\infty} e^{i(2\pi m/T)t} \int_{-T/2}^{T/2} f(t') e^{-i(2\pi m/T)t'} dt'$$

If f(t) is non-periodic its period may be said to be infinite. Put $m/T = \nu$ and $1/T = d\nu$ and let $T \to \infty$. Then, from the definition of the Riemann integral, the sum may be replaced by an integral. Hence

$$f(t) = \int_{-\infty}^{\infty} d\nu \left[e^{i2\pi\nu t} \left\{ \int_{-\infty}^{\infty} f(t') e^{-i2\pi\nu t'} dt' \right\} \right]$$

Let

$$F(\nu) = \int_{-\infty}^{\infty} f(t) e^{-i2\pi\nu t} dt \qquad (A.1)$$

Then it follows that

$$f(t) = \int_{-\infty}^{\infty} F(\nu) e^{i2\pi\nu t} d\nu \qquad (A.2)$$

$F(\nu)$ is known as the Fourier transform of f(t) while f(t) is called the inverse

Fourier transform of $F(\nu)$. If the one is given the other can be obtained from
(A.1) or (A.2) respectively.

A square-wave pulse (p. 125) of duration τ and strength 1 is defined by
the equations

$$f(t) = 1 \qquad |t| < \tau/2$$
$$= 1/2 \quad |t| = \tau/2$$
$$= 0 \qquad |t| > \tau/2$$

It is easy to show by means of (A.1) that the corresponding amplitude
spectrum (Fig. 66) is given by

$$F(\nu) = \tau \,\frac{\sin(\pi\nu\tau)}{(\pi\nu\tau)}$$

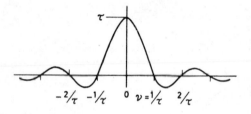

Fig. 66. *Fourier spectrum of a square-wave pulse*

Convolution (section 6.8).

Let $f(t)$ represent the response of a linear system to a unit impulse, t seconds
after the impulse has been applied. A continuous signal, whose strength at a
time λ is $g(\lambda)$, may be considered to consist of an infinite number of
successive very closely spaced impulses, each of strength $g(\lambda)d\lambda$. At a time t,
the system's response to the impulse $g(\lambda)d\lambda$ is evidently $g(\lambda)f(t - \lambda)d\lambda$ since
$t - \lambda$ is the time elapsed after the impulse $g(\lambda)d\lambda$. The total response of the
system at time t is then

$$r(t) = \int_{-\infty}^{\infty} g(\lambda)f(t - \lambda)d\lambda \tag{A.3}$$

$$= \int_{-\infty}^{\infty} g(t - \lambda)f(\lambda)d\lambda \tag{A.4}$$

The integral in (A.3) or (A.4) is known as the convolution integral. The limits $-\infty$ and $+\infty$ are chosen merely for the sake of convenience in analysis. Physically, if $g(\lambda) \equiv 0$ for $\lambda < t_0$, say, and $f(t - \lambda) \equiv 0$ for $t - \lambda < t_1$, that is $\lambda > t - t_1$, the effective limits of integration will be simply t_0 and $t - t_1$.

Taking the Fourier transform of both sides of (A.4),

$$R(\nu) = \int_{-\infty}^{\infty} \int_{-\infty}^{\infty} g(t - \lambda)f(\lambda)e^{-i2\pi\nu t} \, dt d\lambda$$

$$= \int_{-\infty}^{\infty} \left[\int_{-\infty}^{\infty} g(t - \lambda)e^{-i2\pi\nu(t - \lambda)} \, dt \right] f(\lambda)e^{-i2\pi\nu\lambda} \, d\lambda$$

$$= \int_{-\infty}^{\infty} [G(\nu)] f(\lambda)e^{-i2\pi\nu\lambda} \, d\lambda$$

$$= G(\nu) \int_{-\infty}^{\infty} f(\lambda)e^{-i2\pi\nu\lambda} \, d\lambda$$

$$= G(\nu)F(\nu) \tag{A.5}$$

(A.5) shows that the Fourier transform of the convolution of two functions is equal to the product of the Fourier transforms of the functions. The same result follows if we start from (A.3).

Analytic downward continuation (p. 59)
The formal solution of (3.16a) (p. 59) is easily obtained by using the above convolution theorem. Since

$$\frac{1}{r^3} = \frac{1}{\{(x - x_0)^2 + (y - y_0)^2 + z^2\}^{3/2}}$$

the right-hand side of (3.16a) may be regarded as a convolution (in two dimensions) of $1/r^3$ and $(z/2\pi)g(x_0, y_0, z)$. Instead of one frequency as in (A.1) we then have two spatial frequencies ν, μ (that is, wave numbers = reciprocals of wavelengths) in the x and y directions. Taking the two-dimensional Fourier transform of both sides in (3.16a) we get, by the convolution theorem,

$$G(v, \mu) = \frac{z}{2\pi} G_0(v, \mu)e^{-\sqrt{v^2 + \mu^2}\, z} \qquad (A.6)$$

since it can be shown that the Fourier transform of $1/r^3$ is

$$\frac{2\pi}{z} \cdot e^{-\sqrt{v^2 + \mu^2}\, z} \cdot e^{-i2\pi(vx_0 + \mu y_0)}.$$

Therefore

$$G_0(v, \mu) = e^{\sqrt{v^2 + \mu^2}\, z} G(v, \mu)$$

Now, $g(x, y, 0)$ being the given function, $G(v, \mu)$ can be obtained by the numerical evaluation of two double integrals, namely,

$$\int_{-\infty}^{\infty} \int_{-\infty}^{\infty} g(x, y, 0) \frac{\cos}{\sin}(vx + \mu y)dxdy$$

for various values of v and μ. Then $g(x_0, y_0, z)$ is obtained as the inverse Fourier transform of $G_0(v, \mu)$. Towards this end we must numerically evaluate the two double integrals

$$\int_{-\infty}^{\infty} \int_{-\infty}^{\infty} e^{\sqrt{v^2 + \mu^2}\, z} G(v, \mu) \frac{\cos}{\sin}(vx_0 + \mu y_0)dvd\mu$$

for various values of x_0, y_0.

It should be noticed here that the amplitudes of all frequencies in $g(x, y, 0)$ are magnified in downward continuation on account of the exponential factor. The high frequencies (rapid variations) are magnified more quickly (that is, at smaller depths of continuation).

Polynomial expression (section 6.11).

It is evident from the form of (A.1) and (A.2) that for numerical evaluation we can replace a Fourier transform integral by a polynomial in the variable $z = \exp(-i2\pi v \Delta t)$, and the inverse Fourier transform by one in $1/z = \exp(i2\pi v \Delta t)$. Thus, for example,

$$F(v) = \sum_n A_n z^n$$

where the polynomial coefficients A_n are given by the integral of $f(t)$ over the n-th Δt-interval.

References

Where no journal name is given the reference is to *Geophysics* published by the Society of Exploration Geophysics, Tulsa 1, Oklahoma, U.S.A.

1. BATES, L. F. (1951). *Modern Magnetism*. (Cambridge University Press, Cambridge.)
2. GREEN, R. (1960). *Geophys. Prosp.*, **8**, 98.
3. NAGATA, R. (1953). *Rock Magnetism*. (Maruzen and Co., Tokyo.)
4. NIPPOLDT, A. (1930). *Application of Magnetic Measurements to Staking Mining Claims*. (Julius Springer, Berlin.)
5. EVE, A. S. (1932). *Trans. Amer. Inst. min. metall. Engrs*, **97**, 241.
6. KOENIGSBERGER, J. (1928). *Gerlands Beitr. Geophys.*, **19**, 241.
7. CARLHEIM-GYLLENSKÖLD, V. (1910). *A Brief Account of a Magnetic Survey of the Iron Ore Field at Kirunavaara, Stockholm*.
8. SMITH, R. A. (1959). *Geophys. Prosp.*, **7**, 55.
9. ZIETZ, I., and HENDERSON, R. G. (1956), **21**, 794.
10. YÜNGUL, S. (1956), **21**, 433.
11. ZACHOS, K. (1944). *Beitr. angew. Geophys.*, **11**, 1.
12. SMITH, R. A. (1961). *Geophys. Prosp.*, **9**, 399.
13. LEHMANN, H. (1971). *Geophys. Prosp.*, **19**, 133.
14. HEILAND, C. A. (1940). *Geophysical Exploration*, Prentice Hall, New York.
15. HAALCK, H. (Editor) (1953). *Leghrb. d. angew. Geoph.*, Teil l, Berlin.
16. DOBRIN, M. (1952). *Introduction to Geophysical Prospecting*, New York.
17. LACOSTE, L. J. B. (1934). *Physics*, **5**, 178.
18. SIIKARLA, T. (1966). *Geoexploration*, **4**, 139.
19. HAMMER, S. (1939). **4**, 184.
20. GOGUEL, J. (1954). *Geophys. Prosp.*, **2**, suppl. to No. 1.
21. ADLER, J. A. (1942), **7**, 35.
22. NETTLETON, L. L. (1939), **4**, 176.
23. JUNG, K. (1953). *Zeit. f. Geoph.*, p. 54 (Sonderband).
24. PARASNIS, D. S. (1952). *Mon. Not. R. Ast. Soc., Geoph. Suppl.*, **6**, 252.

25. SIEGERT, A. J. F. (1942), **7**, 30.
26. LEGGE, J. A. Jr (1944), **9**, 175.
27. JUNG, K. (1959). *Gerlands Beitr. Geophy.*, **68**, 268.
28. LINDBLAD, A., and MALMQVIST, D. (1938). *Proc. Roy. Swedish Inst. Eng. Res.*, Nr. 146.
29. NABIGHIAN, M. N. (1962). *Geof. Pura e App.*, **53**, 45.
30. LEVINE, S. (1941), **6**, 180.
31. KOLBENHEYER, T. (1968). *Geoexploration*, **6**, 9.
32. SIEGERT, A. J. F. (1942), **7**, 354.
33. ROY, A. (1959). *Geophys. Prosp.*, **7**, 414.
34. TALWANI, M., and EWING, M. (1960), **25**, 203.
35. KELLOGG, O. D. (1953). *Foundations of Potential Theory*. (Dover Publications, New York.)
36. SKEELS, D. C. (1947), **12**, 43.
37. SKEELS, D. C., and WATSON, R. J. (1949), **14**, 133.
38. BULLARD, E. C., and COOPER, R. I. B. (1948). *Proc. roy. Soc.* (A), **194**, 332.
39. JUNG, K. (1937). *Zeit. f. Geoph.*, **13**, 45.
40. JUNG, K. (1953). *Geophys. Prosp.*, **1**, 29.
41. BOTT, M. H. P., and SMITH, R. A. (1958). *Geophys. Prosp.*, **6**, 1.
42. SMITH, R. A. (1959). *Geophys. Prosp.*, **7**, 55.
43. SMITH, R. A. (1960). *Geophys. Prosp.*, **8**, 607.
44. HAMMER, S. (1945), **10**, 50.
45. GRANT, F., and WEST, G. F. (1965). *Interpretation Theory in Applied Geophysics.* (McGraw-Hill.)
46. ELKINS, T. A. (1951), **16**, 29.
47. NETTLETON, L. L. (1954), **19**, 1.
48. ROSENBACH, O. (1959). *Geophys Prosp.*, **7**, 128.
49. HAMMER, S., and ELKINS, T. A. (1938), **3**, 315.
50. COOK, A. H. *et al.* (1951). *Quart. J. geol. Soc. Lond.*, **107**, 287.
51. DAVIS, W. E. *et al.* (1957), **22**, 857.
52. SAXOV, S., and NYGAARD, K. (1953), **18**, 913.
53. GRANT, F. S. (1954), **19**, 23.
54. ROY, A. (1966). *Geoexploration*, **4**, 65.
55a. FAJKLEWICZ, Z. (1959), **24**, **465**.

55b. FAJKLEWICZ, Z. (1965), **30**, 1094.

56. EVE, A. S., and KEYS, D. A. (1956). *Applied Geophysics*, 4th ed. (Cambridge University Press, Cambridge.)

57. FOX, R. W. (1830). *Phil. Trans.,* **130**, 399.

58. SATO, M., and MOONEY, H. M. (1960), **25**, 226.

59. MUSKAT, M., and EVINGER, H. H. (1941), **6**, 397.

60. CARPENTER, E. W., and HABBERJAM, G. M. (1956), **21**, 455.

61. PARASNIS, D. S. (1966). *Mining Geophysics.* (Elsevier Publishing Co., Amsterdam.)

62. HAALCK, H. (Editor) (1953). *Lehrb. d. angew. Geoph.,* Teil 2, Berlin.

63. KING, L. V. (1933). *Proc. roy. Soc.* (A), **139**, 237.

64. SLICHTER, L. B. (1933). *Physics,* **4**, 307.

65. LANGER, R. E. (1933). *Bull. Amer. math. Soc.,* **39**, 814.

66. STEVENSON, A. F. (1934). *Physics,* **5**, 114.

67a. HUMMEL, J. N. (1929). *Zeit. f. Geoph.,* **5**, 89.

67b. HUMMEL, J. N. (1929). *Zeit. f. Geoph.,* **5**, 228.

68. STEFANESCU, S. C., and SCHLUMBERGER, M. (1930). *J. Phys. Radium,* **7**, 132.

69. COMPAGNIE GENERALE DE GEOPHYSIQUE (1955). *Geophys. Prosp.,* **3**, Suppl. 3.

69b. Rijkswaterstaat, The Netherlands (1969). *Standard graphs for resistivity prospecting.* (European Association of Exploration Geophysicists.)

70. MOONEY, H. M., and WETZEL, W. W. (1956). *The Potentials about a Point Electrode.* (University of Minnesota Press, Minneapolis.)

71. FLATHE, H. (1955). *Geophys. Prosp.,* **3**, 95.

72. FLATHE, H. (1955). *Geophys. Prosp.,* **3**, 268.

73. ZOHDY, A. A. R. (1965), **30**, 644.

74. KOEFOED, O. (1960). *Geophys. Prosp.,* **8**, 459.

74b. KOEFOED, O. (1968). *The application of the kernel function in interpreting geoelectrical resistivity measurements.* (Gebrüder Borntraeger.)

75. COLOMBO, U. (1959). *Geophys. Prosp.,* **7**, 91.

76. PEKERIS, C. L. (1940), **5**, 31.

77. BROUGHTON EDGE, A. B., and LABY, T. H. (1931). *The Principles and Practice of Geophysical Prospecting.* (Cambridge University Press.)

78. PARASNIS, D. S. (1970). *Mining and Groundwater Geophysics* (1967) *Proceedings* (Ottawa) 290.

79. LOGN, Ö. (1954), **19**, 739.

80. DE GERY, J. C., and KUNETZ, G. (1956), **21**, 780.

81. MAEDA, K. (1955), **20**, 123.

82. COOK, K. L., and VAN NOSTRAND, R. G. (1954), **19**, 761.

83a. ALFANO, L. (1959). *Geophys. Prosp., 7*, 311.

83b. ALFANO, L. (1960). *Geophys. Prosp., 8*, 576.

83c. ALFANO, L. (1961). *Geophys. Prosp., 9*, 213.

84. VOZOFF, K. (1960), **25**, 1184.

84b. VOZOFF, K. (1970). *Mining and Groundwater Geophysics* (1967) *Proceedings* (Ottawa) 109.

85. BHATTACHARYA, P. K., and PATRA, H. P. (1968). *Direct current electrical sounding.* (Elsevier.)

86. MAILLET, R. (1947), **12**, 529.

87. SCHLUMBERGER, C. and M., and LEONARDON, E. G. (1934). *Trans. Amer. Inst. min. metall. Engrs.*, **110**, 237.

88. HARTSHORN, L. (1925). *J. Inst. elect. Engrs.*, **64**, 1152.

89. BLEIL, D. F. (1953), **18**, 636.

90. MARSHALL, D. J., and MADDEN, T. R. (1959), **24**, 790.

91. VACQUIER, V. *et al.* (1957), **22**, 660.

92. MAYPER, V. (1959). Chap. 10, *Overvoltage Research and Geophysical Applications* (Ed. J. Wait). (Pergamon Press, New York.)

93. BHATTACHARYYA, B. K., and MORRISON, H. F. (1963). *Geophys. Prosp.*, **11**, 176.

94. PARASNIS, D. S. (1965). *Geoexploration, 3*, 1.

95. SUNDBERG, K. (1931). *Gerlands Beitr. Geoph., Ergänzumgs-Hefte*, **1**, 298.

96. SUNDBERG, K., and HEDSTRÖM, H. (1934). *World Petr. Cong. Proc.*, **B(I)**, 107.

97. ZUSCHLAG, T. (1932). *Trans. Amer. Inst. min. metall. Engrs.*, **97**, 144.

98. HEDSTRÖM, H. (1930). *The Oil Weekly*, July 25 and August 8 issues.

99. PARASNIS, D. S. (1971). *Geophys. Prosp.*, **19**, 163.

100. HEDSTRÖM, H. (1937). *Am. Inst. Min. Met. Eng.*, Tech. Publ. 827.

101. ROCHA GOMES, A. A. (1958). *Geophysical Surveys*, (Case histories published by the European Association of Exploration Geophysicists, The Hague), p. 97.

102. MALMQVIST, D. (1965). *Geoexploration, 3*, 175.

103. DEBYE, P. (1909). *Ann. Phys.,* **30**, 57.

104. MARCH. H. W. (1953), **18**, 671.

105.   WAIT, J. R. (1960), **25**, 649.

105a. WAIT, J. R. (1952), **17**, 378.

106a. WESLEY, J. P. (1958), **23**, 128.

106b. WESLEY, J. P. (1958), **23**, 134.

107.   GRAF, A. (1934). *Beitr, angew. Geophys.*, **4**, 1.

108.   SLICHTER, L. B. (1951), **16**, 431.

109.   SLICHTER, L. B., and KNOPOFF, L. (1959), **24**, 77.

110.   PARKHEMENKO, E. (1967). *Electrical properties of rocks.* (Plenum Press.)

111.   PRITCHETT, W. C. (1952), **17**, 193.

112.   COOPER, R. I. B. (1948). *Proc. Phys. Soc. Lond.*, **61**, 40.

113.   YOST, W. J. (1952), **17**, 89.

114.   TUMAN, V. S. (1951), **16**, 102.

115.   BOISSONNAS, E., and LEONARDON, E. G. (1948), **13**, 387.

116.   NIBBLETT, E. R., and SAYN-WITTGENSTEIN, C. (1960), **25**, 998.

117.   CAGNIARD, L. (1953), **18**, 605.

118.   WARD, S. H. (1959), **24**, 761.

119.   RICKER, N. (1940), **5**, 348.

120.   RICKER, N. (1953), **18**, 10.

121.   FAUST, L. Y. (1951), **16**, 192.

122.   HUGHES, D. S., and CROSS, J. H. (1950), **16**, 577.

123.   BAULE, H. (1953). *Geophys. Prosp.*, **1**, 111.

124.   SHUMWAY, G. (1956), **21**, 305.

125.   ANSTEY, N. A. (1957). *Geophys. Prosp.*, **5**, 37.

126.   PARASNIS, D. S. (1966). *Geoexploration*, **4**, 177.

127.   WELIN, E. (1958). *Geophysical Surveys* (Case histories published by European Association of Exploration Geophysicists, The Hague), p. 269.

128.   KREY, T. (1951), **16**, 468.

129.   KREY, T. (1954). *Geophys. Prosp.*, **2**, 61.

130.   LORENZ, G. (1954). *Gerlands Beitr. Geophy.*, **63**, 99.

131.   BAUMGARTE, J. (1955). *Geophys. Prosp.*, **3**, 126.

132.   MENZEL, H., and ROSENBACH, O. (1957). *Geophys. Prosp.*, **5**, 328.

133.   RICKER, N. (1953), **18**, 10.

134.   Symposium (1958). *Geophys. Prosp.*, **6**, 394.

135.   SMITH, M. K. (1958), **23**, 44.

136. SMITH, M. K. (1956), **21**, 337.

137. MUSKAT, M., and MERES, M. W. (1940), **5**, 149.

138. BERRYMAN, L. H. *et al.* (1958), **23**, 223.

139. BORN, W. T. (1941), **6**, 132.

140. DATTA, S. (1968). *Geoexploration,* **6**, 127.

141. ROBINSON, E. (1966). *Geophys. Prosp.,* Suppl. 1.

142. PETERSON, R. A. *et al.* (1955), **20**, 516.

143. Symposium (1960). *Geophys. Prosp.,* **8**, 231.

144. VOGEL, C. B. (1952), **17**, 586.

145. SUMMERS, G. C., and BRODING, R. A. (1952), **17**, 598.

146. MOXHAM, R. M. (1960), **25**, 408.

147. BUDDE, E. (1958), *Geophys. Prosp.,* **6**, 25.

148. HOMILIUS, J., and LORCH, S. (1958). *Geophys. Prosp.,* **5**, 449.

149. HOMILIUS, J., and LORCH, S. (1958), *Geophys. Prosp.,* **6**, 342.

150. HENDERSON, R. G., and ZIETZ, I. (1957), **23**, 887.

151. DE GERY, J. C., and NAUDY, H. (1957). *Geophys. Prosp.,* **5**, 421.

152. HEDSTRÖM, H., and PARASNIS, D. S. (1959). *Geophys. Prosp.,* **7**, 448.

153. TÖRNQVIST, G. (1958). *Geophys. Prosp.,* **6**, 112.

154. HEDSTRÖM, H., and PARASNIS, D. S. (1958). *Geophys. Prosp.,* **6**, 322.

155. BRODING, R. A., and RUMMERFIELD, B. F. (1955), **20**, 841.

156. FEARON, R. E. (1949). *Nucleonics,* **4**, 30.

157. FEARON, R. E. (1949). *Nucleonics,* **4**, 67.

158. KAPPELMEYER, O. (1957). *Geophys. Prosp.,* **5**, 239.

159. GINZBURG, I. I. (1960). *Principles of geochemical prospecting.* (Pergamon Press, New York.)

160. KENDALL, M. G., and MORAN, P. A. P. (1963). *Geometrical Probability.* (Charles Griffin and Co., London.)

161. AGOCS, W. B. (1955), **20**, 871.

162. HAWKES, H. E., and WEBB, J. S. (1962). *Geochemistry in mineral exploration.* (Harper, New York.)

163. WARD, S. H. (1961). *Geophys. Prosp.,* **9**, 191.

164. HALL, D. H. (1959). *J. Geoph. Res.,* **64**, 1945.

165. ÅM, K. (1971). *Geoexploration,* **9**, xxx.

166a. GAY, S. P. (1965), **30**, 818.

166b. GAY, S. P. (1967), *Geophys. Prosp.*, **15**, 236.

167.   HUTCHISON, R. (1958), **23**, 604.

168.   SHARMA, P. V. (1968), **33**, 132.

169a. OGILVY, A. A. *et al.* (1969). *Geophys. Prosp.*, **17**, 36.

169b. BOGOLOVSKY, V. A., and OGILVY, A. A. (1970). *Geophys. Prosp.*, **18**, 261.

169c. BOGOLOVSKY, V. A., and OGILVY, A. A. (1970). *Geophys. Prosp.*, **18**, 758.

170.   LYNCH, E. J. (1962). *Formation Evaluation.* (Harper and Row.)

171.   ANDERS, W. (1966). *Geoexploration*, **4**, 151.

172.   NILSSON, B. (1971). *Geoexploration*, **9**, 35.

173.   AYNARD, C. *et al.* (1961). *Geophys. Prosp.*, **9**, 30.

174.   HALLOF, P. (1967). *IP newsletter cases* xxvii *and* xxviii. (McPhar Geophysics Ltd., Canada.)

175.   NAIR, M. R. *et al.* (1968). *Geoexploration*, **6**, 207.

176.   MORRISON, H. F. *et al.* (1969). *Geophys. Prosp.*, **17**, 82.

177.   DOLAN, W. (1970). *Mining and Groundwater Geophysics* (1967) *Proceedings* (Ottawa), 336.

178.   BHATTACHARYYA, B. K. (1959), **24**, 89.

179.   NABIGHIAN, M. (1970), **35**, 303.

180.   HJELT, S. (1971). *Geoexploration*, **9**, 213.

181.   BEZVODA, V. (1970). *Geophys. Prosp.*, **18**, 343.

182.   DIZIOGLU, M. Y. (1967). *Geoexploration*, **5**, 157.

183.   PORSTENDORFFER, G. (1961). *Geophys. Prosp.*, **9**, 128.

184.   KELLER, G. V. (1971). *Geoexploration*, **9**, 99.

185.   PAAL, G. (1968). *Geoexploration*, **6**, 141.

186.   PATERSON, N. R., and RONKA, V. (1971). *Geoexploration*, **9**, 7.

187.   NORTON, K. A. (1937). *Proc. Inst. Radio Eng.*, **25**, 1203.

188.   GILES, B. F. (1968). *Geophys. Prosp.*, **16**, 21.

189.   PIEUCHOT, M. (1967). *Geophys. Prosp.*, **15**, 7.

190.   MOTA, L. (1954), **19**, 242.

191.   MEIDAV, T. (1960), **25**, 1035.

192.   HABBERJAM, G. (1966). *Geoexploration*, **4**, 219.

193.   BREDE, E. C. *et al.* (1970). *Geophys. Prosp.*, **18**, 581.

194.   D'HOERAENE, J. (1960). *Geophys. Prosp.*, **8**, 389.

195.   KUNETZ, G. (1961). *Geophys. Prosp.*, **9**, 317.

196.   RICE, R. B. (1962), **27**, 4.

197.   AGUILERA, R. *et al.* (1970), **35**, 247.

198.   KREY, Th. (1969). *Geophys. Prosp.,* **17**, 206.

199.   EDELMANN, H. (1966). *Geophys. Prosp.,* **14**, 455.

200.   BARBIER, M., and POULEAU, J. (1970). *Geophys. Prosp.,* **18**, 571.

201.   ZESCHKE, G. (1963). *Econ. Geol.,* **58**, 995.

202.   POLEY, J. Ph., and STEVENINCK, J. van (1970). *Geophys. Prosp.,* **18**, 666.

203.   MAKOWIECKI, L. Z. *et al.* (1970). *Inst. Geol. Sci., Geoph. Pap. No.* 3 (London).

204.   MELTON, B. S. (1971). *Geoph. J.,* **22**, 521.

205.   EVENDEN, B. S., and STONE, D. R. (1971). *Seismic Prospecting Instruments,* vol. 2 (Gebrüder Borntraeger.)

Index